AF792318

Carbon Nanostructures

Series editor

Paulo Araujo, Tuscaloosa, AL, USA

More information about this series at http://www.springer.com/series/8633

Raz Jelinek

Carbon Quantum Dots

Synthesis, Properties and Applications

Springer

Raz Jelinek
Ben-Gurion University of the Negev
Beer Sheva
Israel

ISSN 2191-3005 ISSN 2191-3013 (electronic)
Carbon Nanostructures
ISBN 978-3-319-43909-9 ISBN 978-3-319-43911-2 (eBook)
DOI 10.1007/978-3-319-43911-2

Library of Congress Control Number: 2016947752

© Springer International Publishing Switzerland 2017
This work is subject to copyright. All rights are reserved by the Publisher, whether the whole or part of the material is concerned, specifically the rights of translation, reprinting, reuse of illustrations, recitation, broadcasting, reproduction on microfilms or in any other physical way, and transmission or information storage and retrieval, electronic adaptation, computer software, or by similar or dissimilar methodology now known or hereafter developed.
The use of general descriptive names, registered names, trademarks, service marks, etc. in this publication does not imply, even in the absence of a specific statement, that such names are exempt from the relevant protective laws and regulations and therefore free for general use.
The publisher, the authors and the editors are safe to assume that the advice and information in this book are believed to be true and accurate at the date of publication. Neither the publisher nor the authors or the editors give a warranty, express or implied, with respect to the material contained herein or for any errors or omissions that may have been made.

Printed on acid-free paper

This Springer imprint is published by Springer Nature
The registered company is Springer International Publishing AG Switzerland

Contents

Chapter 1
Introduction

Since their discovery (or one might rather call "first synthesis") approximately 10 years ago, *carbon quantum dots* (first identified as "carbon nanoparticles", but later adopted the term "***Carbon-Dots***", evoking similarities to the widely studied inorganic "quantum dots") and their structurally related cousins *graphene quantum dots* (abbreviated "GQDs") have attracted significant and growing interest in both scientific and technology circles. This interest, also reflected by the almost exponential growth of scientific articles published on the subject (Fig. 1.1, for example, depicts the number of scientific publications having "carbon-dots" mentioned in the title), stems primarily from the unique physical properties of Carbon-Dots, particularly their *luminescent* properties. Carbon-Dots, similar to inorganic fluorescent semiconductor nanoparticles (commonly known as "quantum dots" or "Qdots"), exhibit tunable fluorescence emission, in other words they emit light in different *colors*. This useful property has naturally been a major driving force in the significant expansion of Carbon-Dot research, essentially for the same reasons that Qdots, among the well-known harbingers of the nanotechnology revolution, have attracted huge interest—generating new applications and products in bioimaging, electro-optical and photonic materials, energy harvesting, and others.

Carbon-Dots, however, have an important distinction—they are made of ***carbon***—an abundant and generally nontoxic element, which is also one of the building blocks of life itself. This simple fact makes Carbon-Dots particularly attractive in many applications for which *toxicity risks* pose challenges—from biological imaging to solar cell and photonic device manufacturing. Indeed, biocompatibility has been touted as one of the main advantages of Carbon-Dots in the universe of nanoparticle applications. The carbon composition of Carbon-Dots endows distinctive structural and electronic properties that are different than other nanoparticle families. The carbon-based structural features, however, have also led to difficulties in gaining a fundamental understanding of the photophysical properties of these nanoparticles. Indeed, as will be discussed in the relevant sections below, there is still a comprehensive discussion in the Carbon-Dot community as to the exact mechanisms responsible for the light emission of Carbon-Dots.

© Springer International Publishing Switzerland 2017

R. Jelinek, *Carbon Quantum Dots*, Carbon Nanostructures,
DOI 10.1007/978-3-319-43911-2_1

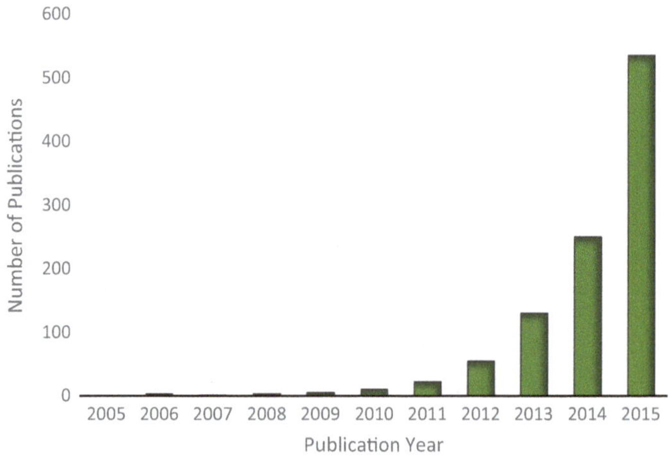

Fig. 1.1 Number of publications with "carbon-dots" mentioned in the title

This Book is thematically focused upon three main themes: synthetic routes for production of Carbon-Dots, elucidating their structural and physical properties, and the diverse applications of Carbon-Dots reported thus far. The book ends with discussion of the future outlook of this field. This treatise is not designed to be an extended scientific "Review" summarizing existing literature, but rather to expose the reader to this exciting research area, varied scientific and technological aspects pertaining to its progress, accomplishments, and challenges. As such, due to the huge volume of publications on Carbon-Dot systems, combined with understandable size and content limitations, only a fraction of published studies are discussed below. It is also likely that by the time this Book will be published, more intriguing and potentially ground-breaking studies of Carbon-Dots will be reported. The reader is thus encouraged to use this book as a starting point for additional journeys in the world of Carbon-Dots. Many excellent reviews have been published and are cited below [1–12].

References

1. F.R. Baptista, S.A. Belhout, S. Giordani, S.J. Quinn, Chem. Soc. Rev. **44**, 4433–4453 (2015)
2. L. Cao, M.J. Meziani, S. Sahu, Y.-P. Sun, Acc. Chem. Res. **46**, 171–180 (2013)
3. A. Cayuela, M.L. Soriano, C. Carrillo-Carrion, M. Valcarcel, Chem. Commun. (Camb.) **52**, 1311–1326 (2016)
4. J.C.G. Esteves da Silva, H.M.R. Gonçalves, Trends Anal. Chem. **30**, 1327–1336 (2011)
5. K.A.S. Fernando, S. Sahu, Y. Liu, W.K. Lewis, E.A. Guliants, A. Jafariyan, P. Wang, C.E. Bunker, Y.-P. Sun, ACS Appl. Mater. Interfaces **7**, 8363–8376 (2015)
6. V. Georgakilas, J.A. Perman, J. Tucek, R. Zboril, Chem. Rev. **115**, 4744–4822 (2015)
7. K. Hola, Y. Zhang, Y. Wang, E.P. Giannelis, R. Zboril, A.L. Rogach, Nano Today **9**, 590–603 (2014)

8. H. Li, Z. Kang, Y. Liu, S.-T. Lee, J. Mater. Chem. **22**, 24230 (2012)
9. S.Y. Lim, W. Shen, Z. Gao, Chem. Soc. Rev. **44**, 362–381 (2015)
10. P.G. Luo, S. Sahu, S.-T. Yang, S.K. Sonkar, J. Wang, H. Wang, G.E. LeCroy, L. Cao, Y.-P. Sun, J. Mater. Chem. B **1**, 2116 (2013)
11. J. Shen, Y. Zhu, X. Yang, C. Li, Chem. Commun. (Camb.) **48**, 3686–3699 (2012)
12. A. Zhao, Z. Chen, C. Zhao, N. Gao, J. Ren, X. Qu, Carbon **85**, 309–327 (2015)

Chapter 2
Carbon-Dot Synthesis

The first report on the production of Carbon-Dots, quite tellingly, featured an almost "accidental" discovery of fluorescent carbon nanoparticles through an unrelated synthetic pathway—a byproduct in arc-discharged soot generated for the synthesis of carbon nanotubes [1]. Intriguingly for the researchers who carried out the experiments, some of the "impurities" identified in the crude nanotube soot were fluorescent (Fig. 2.1), and furthermore different fractions yielded distinct fluorescence wavelengths (e.g., different colors). Analysis of the "fluorescent carbon" as these particles were initially termed indicated that they comprised of very small, nanometer scale, particulates of carbonaceous material, concluded to be small carbon nanotube fragments.

Fluorescent "carbon nanocrystals" were produced, in fact, as side products in other carbon nanotube preparations, and also upon further chemical processing of nanotube samples—both multiwall carbon nanotubes (*MWCNTs*) as well as single-wall nanotubes (*SWNTs*). Figure 2.2, for example, depicts such carbon nanocrystals generated through electrochemical treatment of MWCNTs. In the experiment, Z. Ding and colleagues at the University of Western Ontario, Canada, constructed an electrochemical cell in which one of the electrodes comprised of surface-deposited MWCNTs [2]. Cycling the cell voltage hundreds of times resulted in gradual disintegration of the MWCNT electrode, generating fluorescent Carbon-Dots [still referred to in that work as "carbon nanocrystals"] within the electrolyte solution. The process leading to Carbon-Dot formation was believed to occur through fragmentation of the carbon nanotubes assisted by the electrolyte molecules penetrating into "defect sites" on the nanotubes' surface. The voltage cycling, in particular, was presumed to contribute to breaking C–C bonds, producing hydroxyl and COOH residues. Furthermore, the electrochemical process likely generated hydroxyl and/or oxygen radicals in water which could subsequently react with the nanotubes and promote their cleavage.

While those studies detected Carbon-Dots as a side-product of carbon *nanotubes*, the realization that the "fluorescent carbon" contained new carbon nanostructures has attracted growing interest and led to research aimed at specific synthesis of the new

© Springer International Publishing Switzerland 2017

R. Jelinek, *Carbon Quantum Dots*, Carbon Nanostructures,
DOI 10.1007/978-3-319-43911-2_2

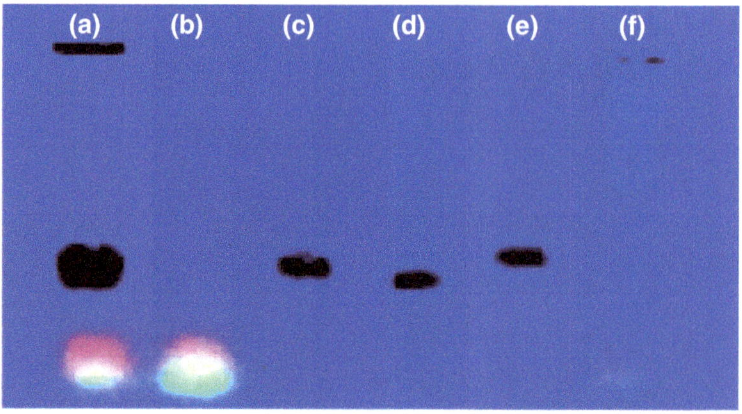

Fig. 2.1 "Fluorescent carbon" nanoparticles identified in agarose gel. The fluorescent fragments were identified in suspensions of single-wall nanotubes (SWNTs). Shown is an agarose gel image under 365 nm illumination. **a** Crude SWNT; a fluorescent band is apparent; **b** purified fluorescent carbon; **c–e** carbon nanotubes; **f** fragmented SWNTs. Reprinted with permission from Xu et al. [27]. Copyright (2004) American Chemical Society

carbon nanoparticles, later termed "carbon quantum dots." Early synthesis procedures mostly used "brute-force" experimental schemes, using high energy impact upon a carbon source thereby generating the fluorescent carbon nanoparticles. Later studies laid the groundwork for more broadly applicable strategies of Carbon-Dot synthesis. Simply put, the stages of many early synthesis approaches for Carbon-Dot production comprised of *fragmentation* of the macroscopic carbon source, inducing (or retaining) the *crystalline organization* of the carbon cores of such (nanoparticle) fragments, and *surface functionalization* of the Carbon-Dots which in most cases has been a prerequisite for the unique photoluminescence properties of the carbonaceous nanoparticles. The diverse Carbon-Dot synthesis schemes discussed below illuminate these aspects.

A representative early example of a brute force approache to construct Carbon-Dots was through *laser ablation* [3]. Laser ablation has been routinely used to produce inorganic nanoparticles from solid substrates; the high energy delivered by the laser beam causes ablation, or removal, of particulates from the target solid substrate, and the substances thus removed have been often found to exhibit nanoscale dimensionalities. This phenomenon was also the case in one of the first direct demonstrations of Carbon-Dots as reported by Y-P Sun and colleagues at Clemson University [3] (in fact, that paper was among the first publications referring to the newly discovered nanoparticles as "carbon-dots"). The researchers prepared a carbon target through high-temperature treatment of a graphite powder/cement mixture and then applied laser ablation (using the widely employed Nd:YAG laser). High temperature (900 °C) was also required to generate the carbon nanoparticles.

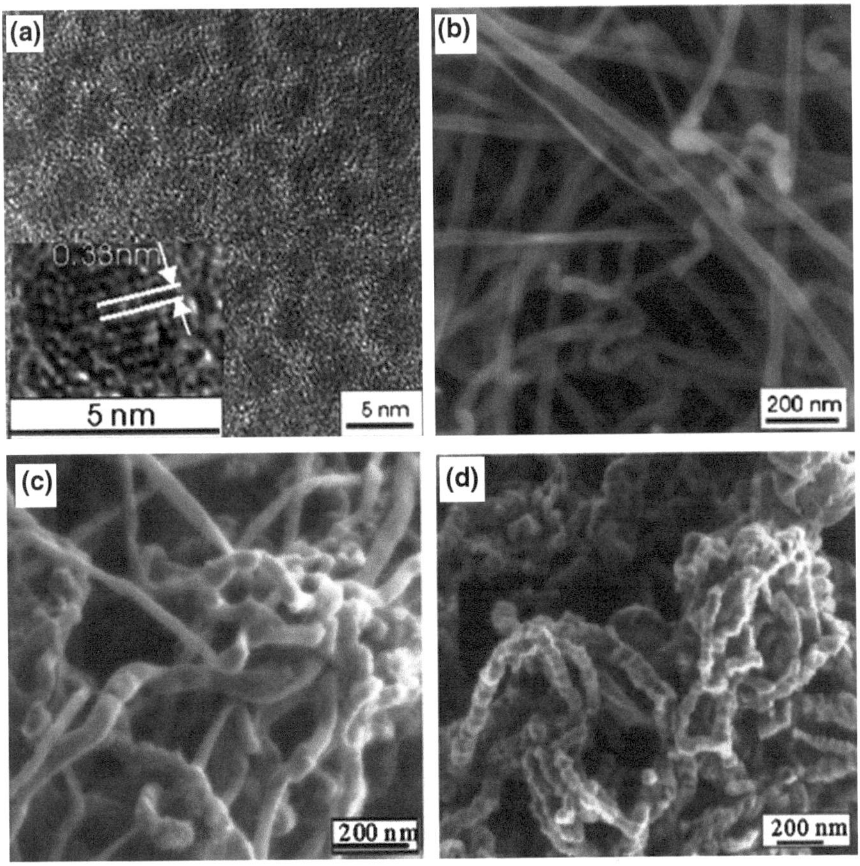

Fig. 2.2 "Carbon nanocrystals" generated upon electrochemical processing of multiwall carbon nanotubes (MWCNTs). **a** High-resolution transmission electron microscopy (HRTEM) of the nanocrystals; **b–d** scanning electron microscopy (SEM) images of the MWCNTs. Pristine sample (**b**), and after 100 voltage cycles (**c**) and 1000 cycles (**d**). The corrugated surface producing the carbon nanocrystals is apparent after voltage cycling. Reprinted with permission from Zhou et al. [2]. Copyright (2007) American Chemical Society

The report by Y-P Sun and colleagues highlights another fundamental aspect of Carbon-Dot synthesis and Carbon-Dot research in general: the realization that Carbon-Dots' *surfaces* have a central role in affecting the fluorescence properties of these nanoparticles. Specifically, the Carbon-Dots described above, produced through laser ablation, were *not fluorescent*, even after purification and treatment with a strong acid. However, as depicted in Fig. 2.3, surface functionalization (or "passivation") with hydrocarbon chains (specifically polyethylene glycol, PEG, in that study) was necessary in order to make the Carbon-Dots photoluminescent. This observation, in fact, crystallizes a fundamental difference between Carbon-Dots and other photoluminescent nanoparticles, particularly semiconductor quantum dots

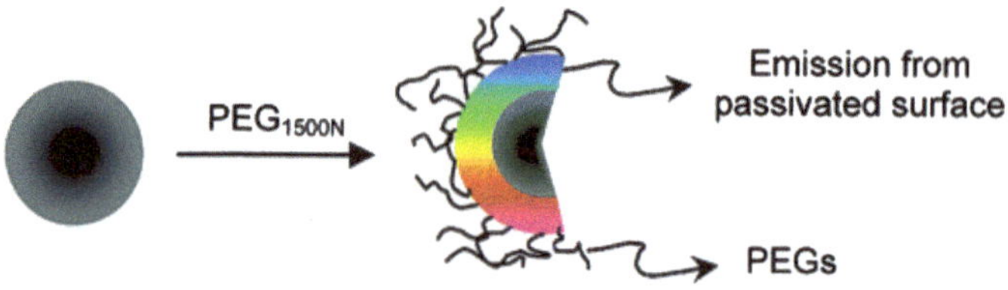

Fig. 2.3 Functionalization of Carbon-Dot surface with polyethylene glycol (PEG). Surface passivation is required to make the Carbon-Dots luminescent. Reprinted with permission from Sun et al. [3]. Copyright (2006) American Chemical Society

(Qdots). In Qdots, the fluorescence emission has been traced to the "quantum effect" in which the nanometer scale of the semiconductor particle dictates coalescence of energy levels allowing excitation of electrons from the valence band into the conducting band in discreet energies. In contrast, a large body of work indicates that the fluorescence properties of Carbon-Dots are associated with the particles' *surface*, specifically "surface defects" responsible for light absorbance in specific wavelengths, i.e., different colors (see detailed discussion in Chap. 3). Accordingly, synthesis schemes for modification of the Carbon-Dots' surface play a central role in the field.

Passivation of Carbon-Dots' surface with varied molecular species other than PEG has been carried out. Y-P Sun and colleagues have observed that other polymers, such as propionylethylenimine-co-ethylenimine (PPEI-EI), generated photoluminescent Carbon-Dots similar to PEG [4]. In fact, surface passivation played an additional important role in the proliferation of Carbon-Dot studies, as it could make the Carbon-Dots *hydrophilic*—thus soluble in water and available for biological applications, particularly bioimaging. Indeed, the varied synthetic routes for coupling functional moieties to Carbon-Dots' surfaces are among the central features contributing to the growth of Carbon-Dot research.

Surface passivation of Carbon-Dots has taken many forms using a variety of molecular constituents. While the significant majority of early studies have employed reactions of Carbon-Dots surfaces with varied organic residues, *inorganic* elements have been examined as well. Y-P Sun and colleagues, for example, showed that doping of the Carbon-Dots with *semiconductor salts* such as ZnS and ZnO prior to surface passivation could enhance the quantum yield of the particles (e.g., their brightness) [5]. Figure 2.4 reveals the formation of the ZnS lattice on the surface of the Carbon-Dots following the doping process, and the dramatic fluorescence of the composite nanoparticles. It should be noted that the luminescence properties in that system were somewhat difficult to explain; the semiconductor salts presumably promoted formation of surface "defects" which contributed to the photoluminescence. While doping with inorganic salts and its effect upon Carbon-Dots' luminescence is interesting from a synthetic standpoint, this line of research has not noticeably progressed because one of the more prominent "selling points" of Carbon-Dots is their production from *organic* building blocks, thus making these nanoparticles supposedly less toxic than inorganic nanoparticles.

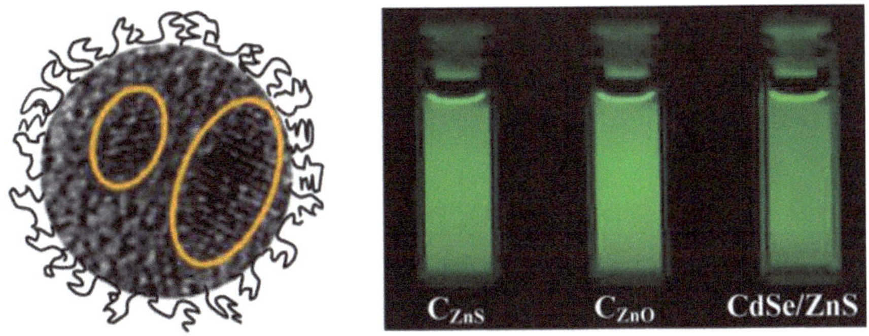

Fig. 2.4 ZnS-doped Carbon-Dots. *Left* Portion of a high resolution TEM (HRTEM) image showing the ZnS lattice planes within the Carbon-Dot (*circled*). Surface functionalization of the Carbon-Dot is illustrated. *Right* photographs of solutions containing the Carbon-Dot/semiconductor composites upon UV irradiation; fluorescence is apparent in all samples. Reprinted with permission from Sun et al. [5]. Copyright (2008) American Chemical Society

Soon after the first reports on production of photoluminescent Carbon-Dots appeared in the literature, a major advance in the synthetic routes for making these intriguing nanoparticles had been achieved. In particular, Carbon-Dot synthesis was demonstrated using readily available precursors, obviating the elaborate preparation processes of the starting materials as described above. Numerous carbon-containing substrates have been successfully utilized as building blocks for construction of Carbon-Dots, including organic solvents, various saccharides, amino acids, proteins, and others. Notably, the underlying feature in most Carbon-Dot preparation schemes is the proposition that formation of the graphitic nanoparticles is thermodynamically (and/or kinetically) favored, regardless of carbon precursor used.

Figure 2.5 depicts a laser-based scheme for production of Carbon-Dots using *toluene*—a common organic solvent—as the carbon precursor. The preparation scheme, demonstrated by X. Li and colleagues at Huazhong University of Science and Technology, China, relied on irradiating the toluene sample with *non-focused* pulsed laser—different than the high power laser irradiation employed in ablation

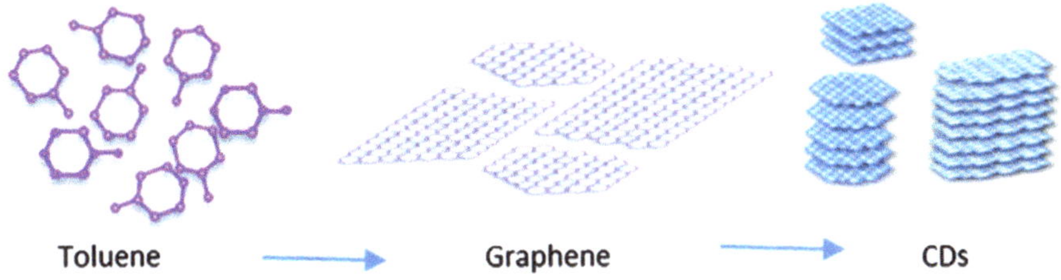

Fig. 2.5 Synthesis of Carbon-Dots using toluene as the starting material. The proposed process consists of formation of graphene sheets through pulse-laser irradiation, followed by fragmentation to yield the Carbon-Dots. Reproduced from Yu et al. [6], with permission of the Royal Society of Chemistry

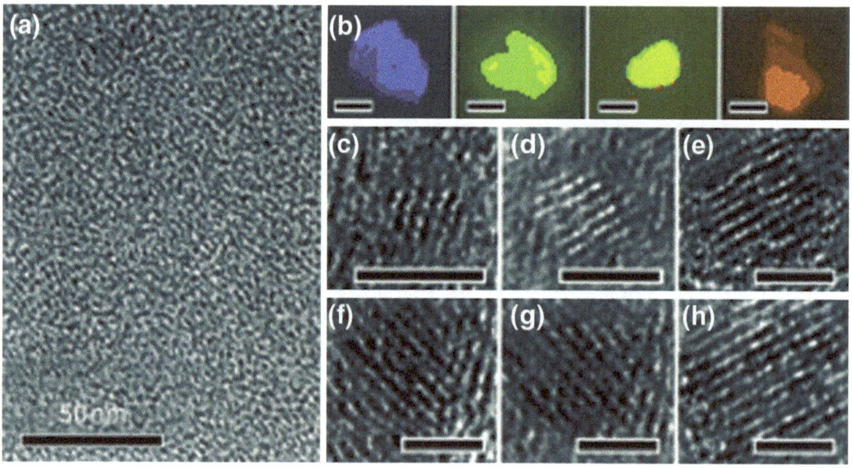

Fig. 2.6 Carbon-Dots produced through an electrochemical method. **a** TEM image of the Carbon-Dots (exhibiting small < 5 nm diameters); **b** photoluminescence images recorded upon excitation at different wavelengths; **c–h** high resolution TEM images showing the crystalline graphite lattice planes. Scale bars correspond to 2 nm. Reprinted from Li et al. [7], with permission from John Wiley & Sons (2010)

processes discussed above [6]. Remarkably, as outlined in Fig. 2.5, irradiation by the non-focused laser beam appeared to have induced transformation of toluene into *graphene* sheaths, which subsequently produced fluorescent Carbon-Dots. The researchers, in fact, achieved certain tuning of the Carbon-Dots *sizes* through modulating the laser power, observing rather surprisingly an *inverse relationship* between laser power and carbon nanoparticles' sizes.

Electrochemical methodologies have been also demonstrated for Carbon-Dot synthesis. S.-T. Lee and colleagues at Hong Kong S.A.R., for example, utilized graphite electrodes as carbon source [7]. Specifically, the researchers employed graphite rods as both cathode and anode in an electrochemical cell setup, with NaOH/ethanol as the electrolyte solution. Passage of current through the electrochemical circuit resulted in "chipping" of the graphite rods, producing Carbon-Dots exhibiting different colors (i.e., luminescence wavelengths). While the Carbon-Dots generated in this experiment *did not* possess uniform spherical morphologies (Fig. 2.6), distinct luminescence colors were observed, ascribed to variations in particle sizes and surface defects. The high-resolution TEM images in Fig. 2.6 confirm that the nanoparticles could indeed be referred to as "Carbon-Dots," as they exhibit the typical graphitic crystalline cores.

Electrochemistry has been implemented by other researchers to produce Carbon-Dots since the technique is easy to carry out and the product yield is usually high. Figure 2.7 presents an experimental setup of an electrochemical cell designed to utilize a graphite electrode as a carbon source for Carbon-Dots, and illustrates the visual progression of the reaction. This electrochemical cell approach, developed by K.P. Loh and colleagues at the National University of Singapore, employed an *ionic*

Fig. 2.7 Production of Carbon-Dots through exfoliation of graphite in ionic liquids. *Top* photographs showing the progression of the electrochemical reaction in which the graphite anode is placed in an electrolyte solution comprising the ionic liquid 1-butyl-3-methylimidazolium tetrafluoroborate ([BMIm][BF_4]). The *brownish* color corresponds to carbon nanoparticles. Panel **f** shows the appearance of the graphite electrode after the reaction. *Bottom* an illustration of the proposed mechanism: the ions (originated from the ionic liquid) penetrate between the graphite sheets, generating carbon nanotubes and Carbon-Dots. Reprinted with permission from Lu et al. [8]. Copyright (2009) American Chemical Society

liquid (*IL*) as the electrolyte [8]. Ionic liquids have been used as "green" substitutes for organic solvents and could possess powerful solvation capabilities. In the context of the Carbon-Dot fabrication process described in Fig. 2.7, the ionic liquids, specifically the positive *imidazolium ion*, had dual roles. First, the imidazolium ions could transform into reactive *radicals* upon accepting electrons from the anode in the electrochemical cell; the radicals reacted with, and induced "fraying" of the graphite sheets, giving rise to the small carbon fragments. Second, as shown in Fig. 2.7b, the ions inserted into the graphite layers, thereby enhancing the exfoliation process. An apparent limitation, however, of many electrochemistry-based Carbon-Dot synthesis schemes such as the experiment in Fig. 2.7 has been the rather broad distribution of nanoparticle sizes and morphologies, necessitating further separation and purification steps.

Numerous other synthetic routes have utilized "wet chemistry" to construct Carbon-Dots. Figure 2.8 outlines a typical solution-based strategy for Carbon-Dot synthesis. The process, developed by R. Liu and colleagues at the Max-Planck-Institut fur Polymerforschung, Germany, was based upon embedding the carbon source (phenol/formaldehyde resin) within porous silica colloids acting as host matrix [9]. A high-temperature carbonization process (e.g., pyrolysis) generated Carbon-Dots which were subsequently released through dissolution of the silica scaffold; further surface passivation of the dots gave rise to the typical luminescence properties. While pyrolysis was the main driving force for Carbon-Dot production, the porous silica carriers were instrumental in preventing

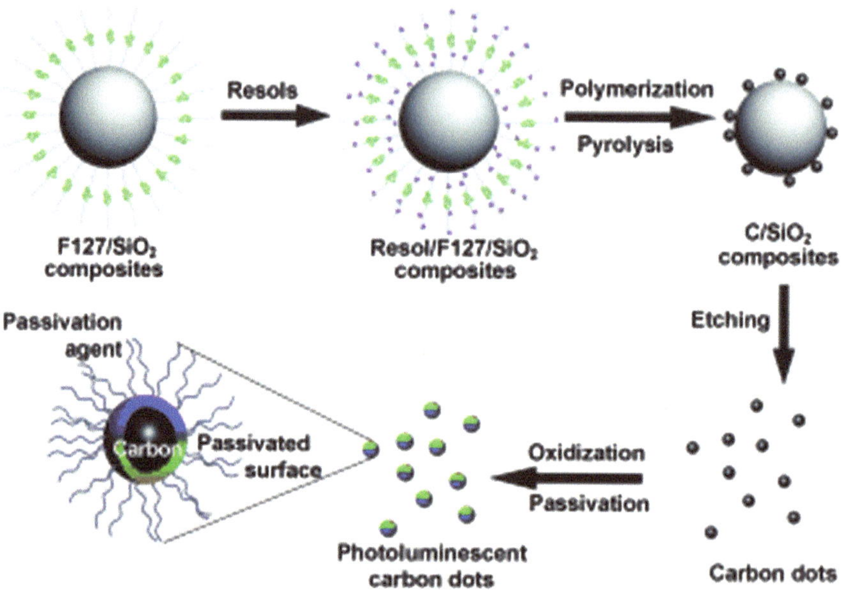

Fig. 2.8 Synthesis of Carbon-Dots within a silica matrix. Reprinted from Liu et al. [9], with permission from John Wiley & Sons (2009)

aggregation of the nanoparticles. It is notable that the synthesis scheme in Fig. 2.8 presumably yielded *amorphous* Carbon-Dots rather than Carbon-Dots comprising *crystalline* graphitic cores, albeit exhibiting strong, multicolor luminescence. This result lends support to the proposed relationship between the luminescence properties of Carbon-Dots and putative *surface defects* (see Chap. 3).

As expected in a young and evolving scientific field, proliferation of studies led to development of rather simple synthesis schemes. In particular, Carbon-Dots can be readily produced even without the use of structure-directing matrixes such as the colloidal silica host (e.g., Fig. 2.8). Carbon *soot*, generated in numerous carbon-processing reactions, has been found to be a useful precursor substance for Carbon-Dots. S.C. Ray and colleagues at the Indian Institute for the Cultivation of Science have shown that soot produced by candle burning could be a source of Carbon-Dots, generated simply through refluxing the soot in a strong acid (such as nitric acid), followed by size separation of the smaller nanoparticles (Fig. 2.9) [10]. The strong oxidation properties of the acid were crucial for dissolution of large carbon aggregates in the soot; the acid further reacted with the carbon colloids to produce the oxygen- and nitrogen-containing surface defects associated with the fluorescence emission of the particles. Such soot-based techniques, however, yielded Carbon-Dots exhibiting relatively low brightness (e.g., low quantum yield), likely because of the large particles produced and the broad size distribution. Other soot-based methods for production of Carbon-Dots were introduced, employing

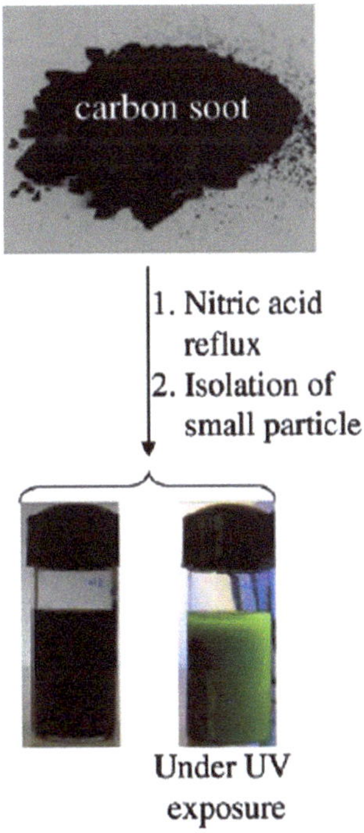

Fig. 2.9 Preparation of Carbon-Dots by acid treatment of carbon soot. Candle-generated soot was refluxed in a strong acid, producing Carbon-Dots. Following purification, the Carbon-Dot solution exhibited fluorescence (bottom photograph). Reprinted with permission from Ray et al. [9]. Copyright (2009) American Chemical Society

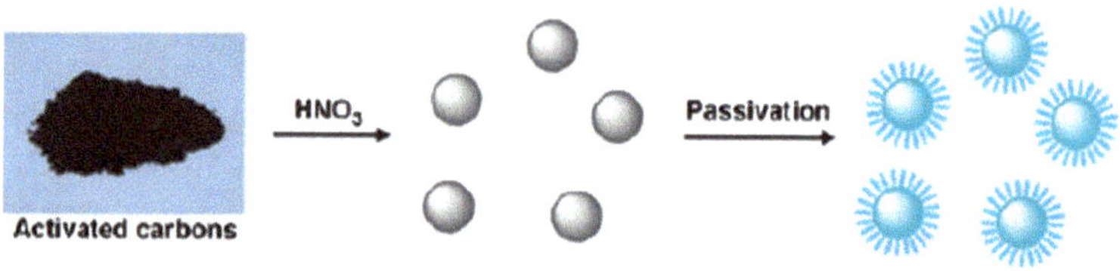

Fig. 2.10 Synthesis of Carbon-Dots from activated carbon. Reproduced from Qiao et al. [12], with permission of the Royal Society of Chemistry

acid treatment of soot generated from varied carbon reactions like combustion of natural gas [11].

Figure 2.10 outlines another procedure for production of Carbon-Dots using a natural carbon source [12]. In that important work, Q. Huo and colleagues at Jilin University, China, successfully prepared Carbon-Dots through acid-induced oxidation of *activated carbon* (i.e., carbon from natural sources such as wood, which following physical treatment—usually application of hot gas—becomes highly porous or "active"). After purification of the Carbon-Dots produced by the chemical treatment, surface passivation was carried out using different amine-terminated organic agents yielding water-soluble luminescent Carbon-Dots.

The study outlined in Fig. 2.10 represents an important leap in Carbon-Dot research, since it showed that the carbon nanoparticles could be produced (in large quantities) from readily available reagents through easily applicable synthetic schemes and with no need for sophisticated and/or expensive instrumentation. It is safe to say that the synthetic progress accomplished by Q. Huo and others constituted a major driving force for the significant growth of activity and applications of Carbon-Dots.

In parallel with the proliferation of "chemical oxidation" techniques starting around 2010 for production of Carbon-Dots [generally involving reaction of carbonaceous substances with strongly oxidizing acids] other solution-based Carbon-Dot synthesis strategies have been developed. In particular, *hydrothermal treatment* has likely become the most common procedure for production of the Carbon-Dots. This generic approach is quite striking in its simplicity and versatility, generating Carbon-Dots from varied carbon sources. The crux of hydrothermal synthesis of Carbon-Dots is a process in which high temperature induces condensation of the carbonaceous building blocks and crystallization of the graphitic core. Another important feature of hydrothermal synthesis schemes has been the observation that residues from the carbon reagents are still retained upon the *surface* of the crystalline cores, thereby bestowing the nanoparticles varied, and chemically controllable, functionalities and optical properties.

Figure 2.11 illustrates a proposed mechanism for assembly of Carbon-Dots through hydrothermal treatment [13]. This particular process, developed by S.P. Lau and colleagues at the Polytechnic University of Hong Kong, was carried out via microwave heating of glucose as the carbon source. According to the experimental model (Fig. 2.11), the Carbon-Dots were assembled through

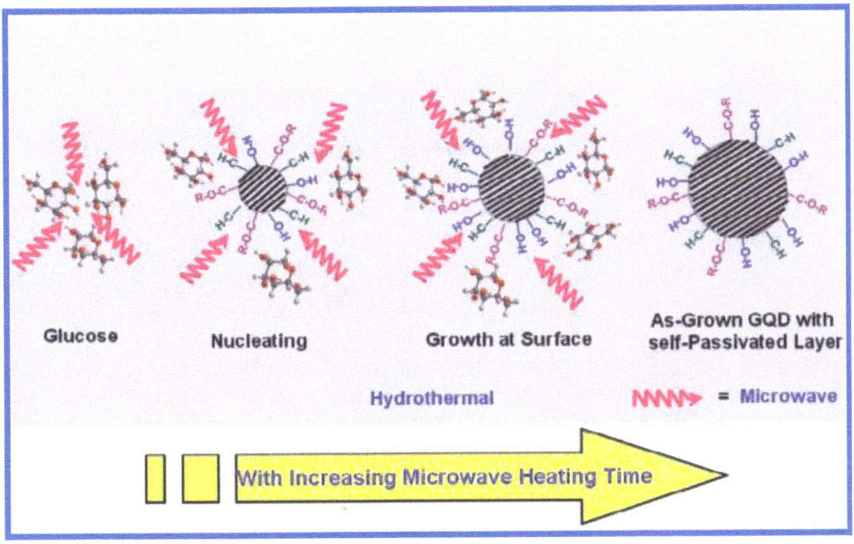

Fig. 2.11 Hydrothermal synthesis of Carbon-Dots using glucose as a carbon source. A model for nucleation and growth of Carbon-Dots induced by microwave heating of glucose solution. Reprinted with permission from Tang et al. [13]. Copyright (2006) American Chemical Society

a nucleation process, leading to gradual growth of the crystalline nanoparticle core and formation of a "self-passivated" layer comprising functional fragments of the glucose building blocks. Intriguingly, these microwave-generated Carbon-Dots featured excitation/emission in the deep blue spectral range, a rather unique property (as the large majority of reported Carbon-Dots emit at longer wavelengths).

Microwave-based synthesis has been employed in other Carbon-Dot fabrication routes. Figure 2.12 illustrates a scheme for preparing Carbon-Dots doped with nitrogen atoms (e.g., N-doped Carbon-Dots) through microwave treatment of calcium citrate and urea. Notably, this simple synthetic route, developed by Y. Zhang and colleagues at Shanghai Jiao Tong University, China, presented a *heterogeneous* synthesis process utilizing a mixture of a *powder* (calcium citrate) and *solution* (urea) [14]. The resultant Carbon-Dots emitted high luminescence both in solution and in the solid phase, a somewhat unique property in Carbon-Dot systems (as Carbon-Dots generally exhibit fluorescence *quenching* in solid phases).

High-temperature carbonization for fabricating Carbon-Dots has been carried out in solvents other than water. M. Kreiter, C-Y. Liu, and colleagues at the Max-Planck-Institut for Polymerforschung showed that heat treatment of carbon sources such as citric acid dispersed within *noncoordinating solvents* [i.e., non-polar, organic solvents] produced highly luminescent Carbon-Dots [15]. The approach was actually inspired by synthesis pathways developed for production of *inorganic quantum dots*, in which nonpolar solvents were believed to promote assembly and transformation of the nanoparticles' building blocks into crystalline

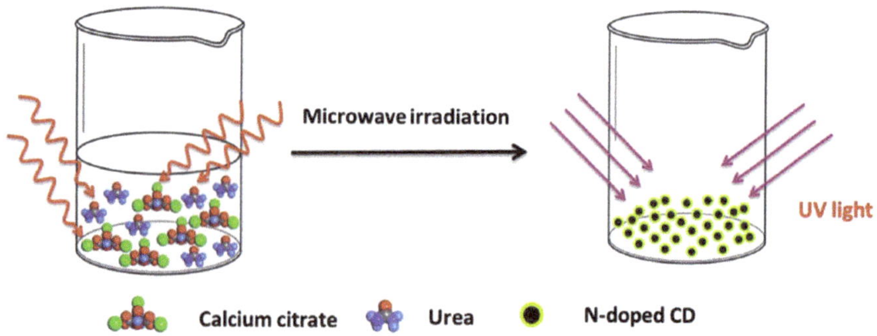

Fig. 2.12 Generation of N-doped Carbon-Dots through microwave heating. Synthesis was carried out upon microwave irradiation of a heterogeneous mixture of calcium citrate (powder) and urea (solubilized). Reproduced from Xu et al. [14], with permission of the Royal Society of Chemistry

colloids. In case of the Carbon-Dots, the simple one-step synthesis scheme reported by Kreiter and his colleagues was particularly successful in increasing the luminescence (i.e., quantum yield) of the particles, an observation that might be related to more effective surface passivation by amine derivatives co-added to the reaction mixture; in that regard, the nonpolar solvents likely enhanced the reactivity between the Carbon-Dots and the amine residues.

Several synthetic pathways have led to fabrication of Carbon-Dots exhibiting somewhat unusual luminescence properties. H. Lin and colleagues at the Chinese Academy of Sciences synthesized N-doped Carbon-Dots via a simple procedure involving triaminobenzene as the carbon source (Fig. 2.13) [16]. The resultant Carbon-Dots contained a relatively high abundance of nitrogen-containing functional groups, including amines, amides, and other carbon-nitrogen units. Intriguingly, these N-doped Carbon-Dots exhibited a strong emission in *yellow*, i.e., around 600 nm—a rare observation as the majority of Carbon-Dots reported thus far had maximal emission in shorter wavelengths (i.e., green-blue). Similarly, in contrast to other studies of Carbon-Dots, no excitation-dependent emission was apparent. These distinct photophysical properties most likely reflect modulation of the Carbon-Dots' surface states of by the nitrogen-containing functional groups (via mechanisms that are still unresolved).

Fig. 2.13 Yellow Carbon-Dots from triaminobenzene. The simple synthesis scheme produced Carbon-Dots emitting in *yellow*. Reprinted with permission from Jiang et al. [16]. Copyright (2015) American Chemical Society

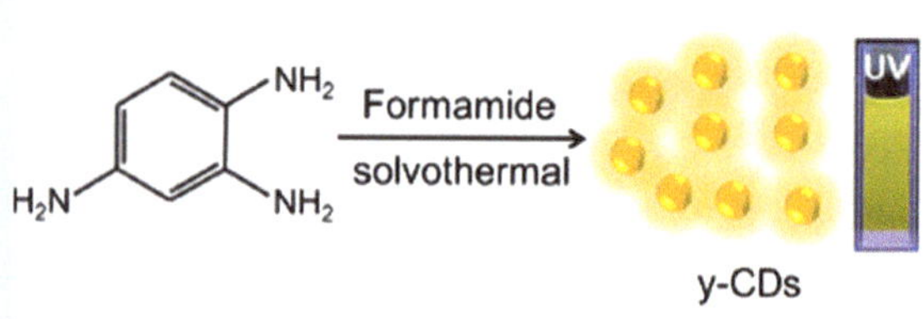

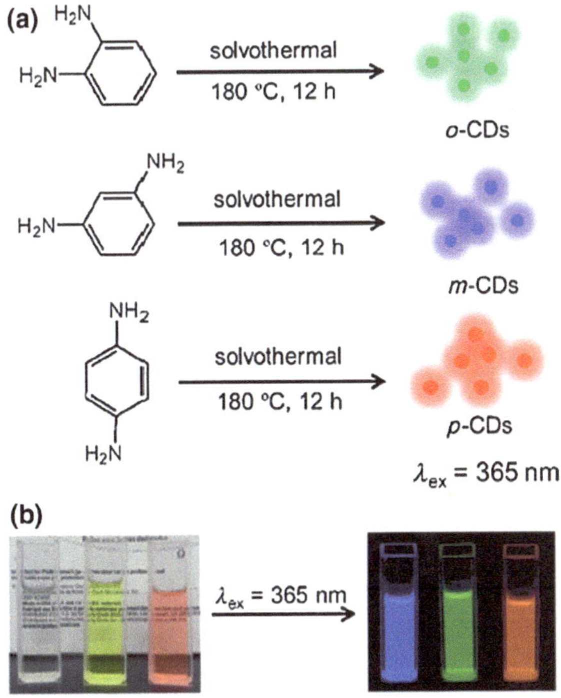

Fig. 2.14 Carbon-Dots emitting different colors from phenylenediamine isomers. **a** Specific colors of Carbon-Dots generated from the *ortho*, *meta*, and *para* isomers. **b** Photographs showing the solution colors in regular light (*left*) and upon UV irradiation (*right*). Reprinted from Jiang et al. [28], with permission from John Wiley & Sons (2015)

That same research group has shown that different *isomers* of phenylenediamine produced Carbon-Dots displaying distinct colors (Fig. 2.14) [16]. Remarkably, as shown in Fig. 2.14, the three isomers *para*-phenylenediamine, *ortho*-phenylenediamine, and *meta*-phenylenediamine, gave rise to the three "basic" colors red, green, and blue (i.e., "RGB"), upon excitation by UV light. This observation was presumably related to the different nitrogen contents (and sizes) of the carbon nanoparticles. Importantly, the full RGB range generated by the Carbon-Dots could be exploited for the production of films exhibiting different predesigned colors by doping polymer films with specific Carbon-Dot compositions.

Numerous methods have been devised to prepare Carbon-Dots from *natural sources*, and it is fair to say that the huge variety of carbon-containing substances will continue to inspire researchers to develop new synthesis schemes using common and readily available carbon materials. Figure 2.15 illustrates an experiment in which *waste frying oil* was used to produce Carbon-Dots [17]. In fact, the simple synthetic scheme, demonstrated by J.S. Yu and colleagues at Nanjing University, China, generated Carbon-Dots that were also doped with *sulfur* atoms. The S-doped Carbon-Dots were used for cell imaging, and their luminescence properties closely depended upon the solution pH, presumably affecting the degree of protonation of the carboxyl residues upon the Carbon-Dots' surface. In a broader context, the system depicted in Fig. 2.15 is an example of the widely-used synthetic pathways for Carbon-Dot *doping* using different precursors.

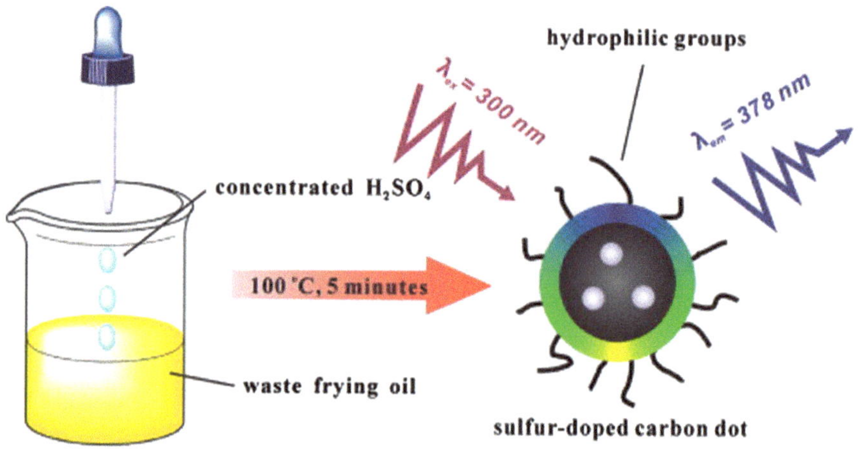

Fig. 2.15 Preparation of sulfur-doped Carbon-Dots from waste frying oil. Reprinted from Hu et al. [17], Copyright (2014), with permission from Elsevier

Sugars have been obvious candidates for Carbon-Dot synthesis and indeed many reports have described synthetic routes utilizing basic sugars, such as glucose, fructose, and others. However, the use of building blocks other than simple hydrocarbons has yielded in many instances Carbon-Dots exhibiting more interesting properties compared to simple sugar units. J. Tang and colleagues at the Chinese Academy of Sciences, for example, synthesized Carbon-Dots from *bovine serum albumin* (*BSA*), a widely used and commercially available protein [18]. The one-pot synthesis scheme involved hydrothermal treatment of the protein in the presence of an amine derivative (serving as the surface passivation agent); the pyrolysis reaction initiated by the high temperature gave rise to Carbon-Dot formation, exhibiting bright blue emission. This simple process also highlights the use of *proteins* as potential carbon sources in Carbon-Dot synthesis. The wide variety of proteins, availability in large quantities in many instances, and water solubility (for most proteins) point to their use as versatile starting materials in Carbon-Dot synthesis.

Figure 2.16 presents a straightforward strategy for construction of Carbon-Dots from *folic acid* as the carbon source [19]. The choice of folic acid in that study aimed to make the Carbon-Dots selective toward the *folate receptor* which is overexpressed in many cancer cells. The Carbon-Dots were synthesized in a single-step hydrothermal process without the need to first synthesize Carbon-Dots and subsequently conjugating the as-prepared Carbon-Dots with folic acid. Indeed, the folic acid-derived Carbon-Dots could selectively stain cancer cells through honing onto the folate receptors on the cell surfaces. This study represents the seemingly endless variability of Carbon-Dot synthesis pathways, reflecting the realization that any carbon-containing compound might be used in practice to produce Carbon-Dots.

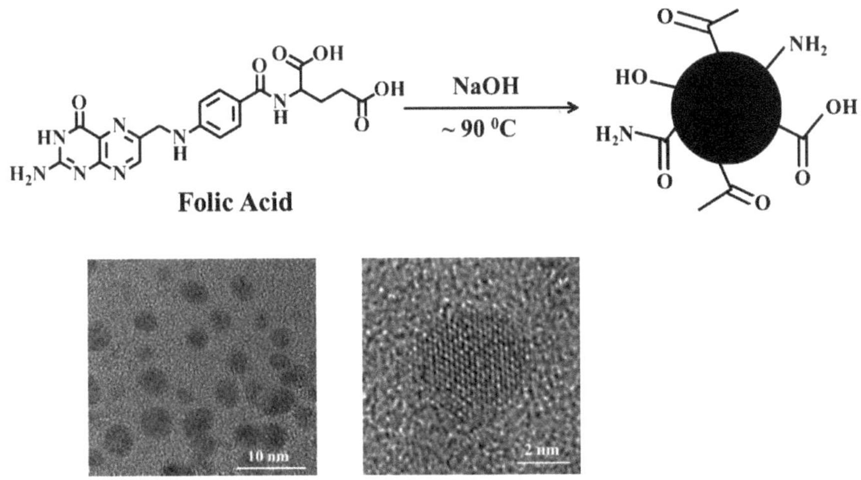

Fig. 2.16 Carbon-Dot preparation from folic acid. *Top* Scheme of the thermal process for generation of the Carbon-Dots in basic aqueous solution. *Bottom* High-resolution TEM of the Carbon-Dots showing the crystalline graphitic core structure. Adapted from Bhunia et al. [19]

In some instances, the carbonaceous reagents employed for synthesis of the Carbon-Dots had multiple functions—not only serving as the carbon source. R. Zboril and colleagues at Olomuc University, Czech Republic, have used *surfactants* as the starting materials in Carbon-Dot synthesis [20] The synthetic route comprised of hydrothermal treatment of a commonly used surfactant, cetylpyridinium chloride monohydrate (Fig. 2.17), which yielded relatively uniform Carbon-Dot samples. Intriguingly, the luminescence of the Carbon-Dots was found to be sensitive to *solvent polarity*, producing different solvent-dependent colors upon UV excitation (Fig. 2.17, bottom). The surfactants in that Carbon-Dot system served not only as the carbon source, but also as stabilizers of the carbon nanoparticles, promoting their solubilization in different solvents, and affecting the "surface states" responsible for the luminescence properties of the particles.

Among the varied synthesis schemes for production of Carbon-Dots, some techniques stand out as seemingly serendipitous. This description fits the strategy depicted in Fig. 2.18, in which Carbon-Dots were prepared by mixing two ubiquitous chemicals—acetone and sodium hydroxide—and incubating the mixture for a few days [21]. The experiment, carried out by X. Ji and colleagues at South Central University, China, unexpectedly generated Carbon-Dots, presumably through the well-known aldol reaction, undergone in the case here by acetone in the alkaline solution environment (Fig. 2.18a). The unsaturated ketone formed by the reaction might have further polymerized, producing oligomeric chains which condensed into the Carbon-Dots. While the researchers reported that the Carbon-Dots exhibited the typical excitation-dependent photoluminescence spectra, the carbon cores of the nanoparticles appeared *amorphous* rather than the usual crystalline organization. Another interesting feature of the system was the formation

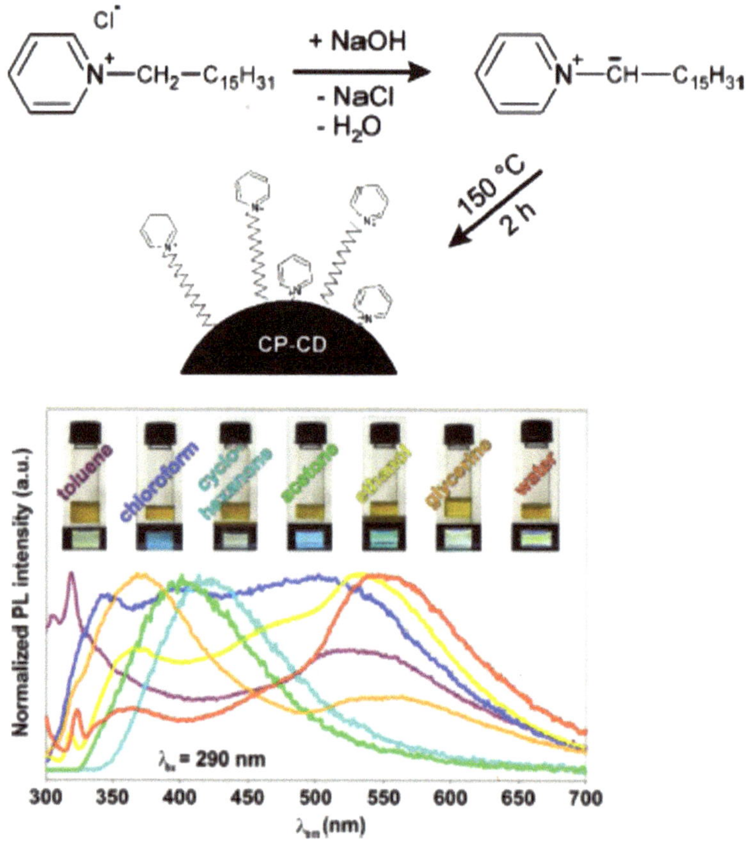

Fig. 2.17 Surfactant-generated Carbon-Dots exhibiting solvent-dependent colors. *Top* Scheme of the Carbon-Dot synthesis starting from cetylpyridinium chloride monohydrate; *bottom* Carbon-Dot solution colors and emission spectra in the indicated solvents. Reprinted with permission from Kozák et al. [20]. Copyright (2013) American Chemical Society

of a *porous carbon network* following high-temperature (800 °C) calcination of the Carbon-Dots, which served as an excellent storage medium in sodium ion batteries.

In parallel with the rapidly proliferating reports on *Carbon-Dots*, many studies have focused on ***graphene quantum dots*** (***GQDs***) (Fig. 2.19). While GQDs have been perceived as distinct morphological entities, their structural and photophysical properties are in many respects similar to those of Carbon-Dots, reflecting the nanoscale dimensions, crystalline organization of the graphitic cores, and surface modification as a primary factor affecting the luminescence wavelength. The main distinction between GQDs and Carbon-Dots concerns the overall shape of the nanoparticles; Carbon-Dots likely constitute isotropic [spherical] nanoparticles, while GQDs are believed to adopt "disk-like" structures, comprising few graphite/graphene layers (i.e., thickness of a few nanometers) (Fig. 2.19).

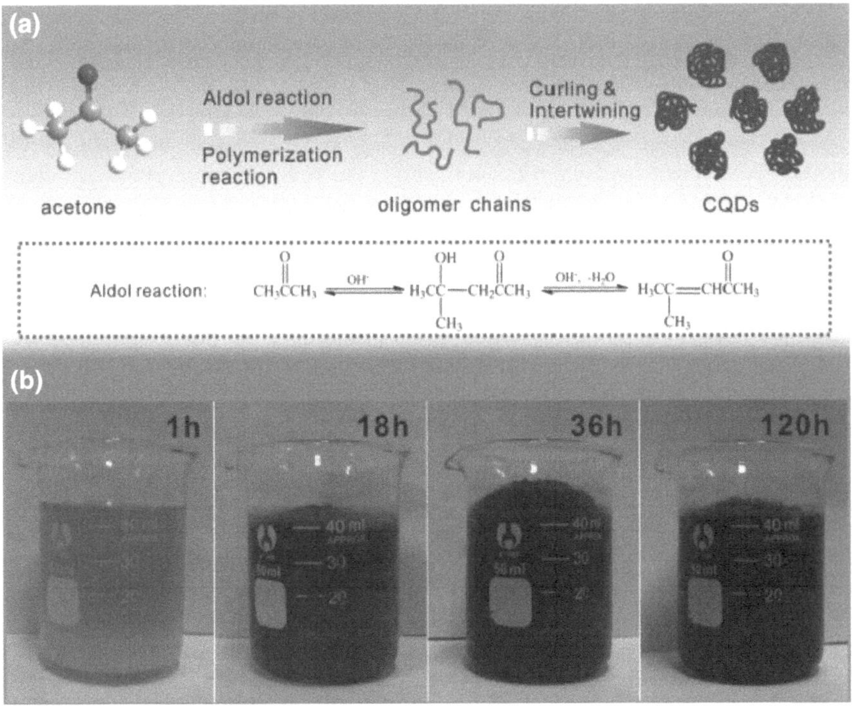

Fig. 2.18 Carbon-Dot synthesis from acetone in NaOH solution. **a** Proposed scheme for synthesis of the Carbon-Dots through the aldol reaction of acetone; **b** the reaction mixture after different incubation times. Formation of the condensed Carbon-Dot suspension is apparent after few days. Reprinted from Hou et al. [21], with permission from John Wiley & Sons (2015)

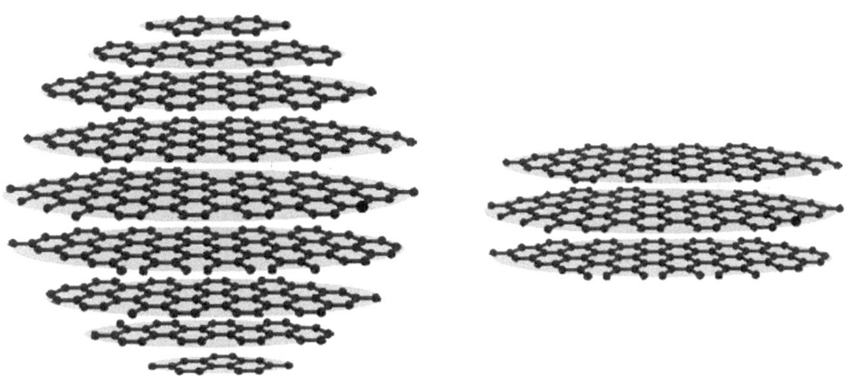

Fig. 2.19 Structural features of carbon-dots and graphene quantum dots. Generic structures of Carbon-Dots (*left*) and GQDs (*right*). Carbon-Dots are generally perceived as spherical aggregates, while GQDs comprise fragments of graphene layers

While some researchers consider GQDs a "special case" of Carbon-Dots, their fabrication and applications have followed, in fact, somewhat different paths. The first reports on GQDs referred to the "dots" as nanometer-scale domains of two-dimensional graphene fabricated through nanolithography (for example high-resolution electron beam lithography [22]). Several studies highlight the difference between the mostly *top-down* methods (such as lithography) for fabrication of GQDs, in comparison with *bottom-up* chemical synthesis techniques employed in the case of Carbon-Dots [and specifically discussed in this Book]. Obviously, precise particle dimensions that can be obtained via lithography have allowed investigation of the unique physical phenomena associated with GQDs.

GQDs have been fabricated also via "wet chemistry" methods similar to techniques described above in case of Carbon-Dot synthesis. K. Mullen and colleagues at the Max-Planck-Institut fur Polymerforschung, Germany, reported synthesis of discoidal GQDs using a coronene derivative as the starting material [and carbon source] (Fig. 2.20). The preparation method involved initial self-assembly of the coronene in solution, followed by high-temperature transformation (pyrolysis) which generated graphite-like aggregates; subsequent exfoliation-through-oxidation and surface functionalization of the reaction products yielded GQDs exhibiting thickness of 2–3 nm (corresponding to 2–3 graphene layers) and diameters of tens

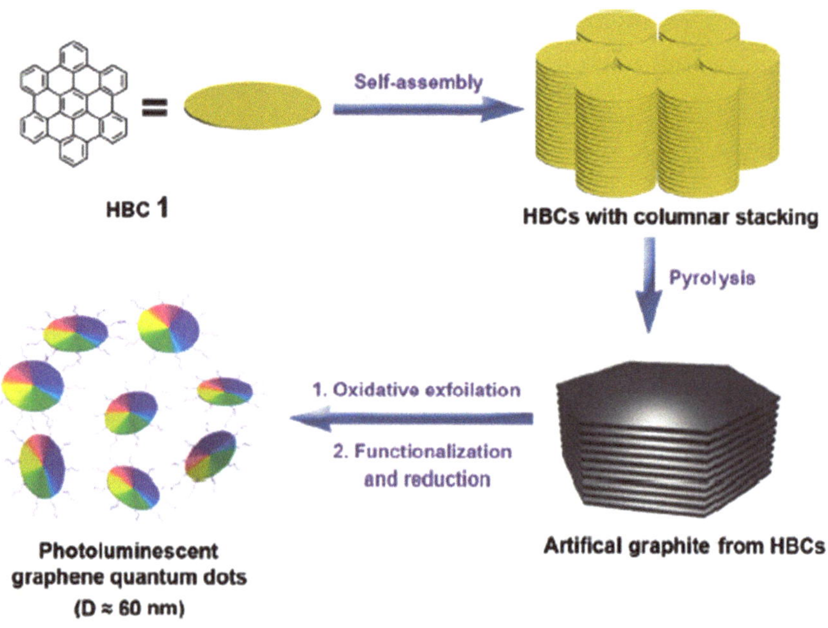

Fig. 2.20 Graphene quantum dots from coronene as the starting material. A scheme depicting the bottom-up synthesis process of the GQDs, likely originating from pyrolysis of the stacked layers of hexabenzocoronene (HBC). Reprinted with permission from Liu et al. [29]. Copyright (2011) American Chemical Society

of nanometers. The formation of GQDs in this case was attributed to the anisotropic rigid structure of the coronene building block, leading to assembly of the initial stacked columns that were subsequently fragmented, forming the GQDs (Fig. 2.20). It should be noted that the synthetic procedure yielded surprisingly uniform dispersion of the GQDs, highlighting the potential of "bottom-up" synthetic techniques to produce anisotropic carbon nanostructures in a controlled manner.

The study depicted above in Fig. 2.20 utilized a specific organic molecule for assembly of organized structures, ultimately generating the crystalline carbon cores of the GQDs. In many instances, however, GQDs have been synthesized from widely used carbon materials that already exhibit supramolecular ordering, such as carbon nanotubes or graphite "flakes". Figure 2.21, for example, depicts a synthetic pathway in which carbonaceous assemblies such as multiwall carbon nanotubes (MWCNTs) or graphite fragments were reacted with potassium atoms to yield GQDs [23]. The study, carried out by S. Zhang and colleagues at Sheffield University, UK, revealed intercalation of the potassium atoms within the layered structures of the MWCNTs and graphite-promoted exfoliation and disintegration upon exposure to air [due to the high reactivity of the electron-donating potassium]. The fragmentation and oxidation processes associated with the combustion reaction between potassium and oxygen gave rise to GQDs comprising mostly a *single* graphite monolayer in relatively high yields.

Many published schemes for synthesis of GQDs employed *graphene oxide* (*GO*) as the starting material. Such strategies have taken advantage of the layered sheet organization of GO, obviating the need for initial assembly processes such as presented above (e.g., Figs. 2.20 and 2.21). Indeed, the use of building blocks which already employ graphene layers distinguishes *GQDs* from *Carbon-Dots*, as GQDs generally retain the few-layered organization of graphene. In such GO-based techniques, GQDs are essentially produced through disintegration/fragmentation of the GO sheets by varied experimental means, complemented by surface functionalization to generate photoluminescence. An example of such a GQD synthetic

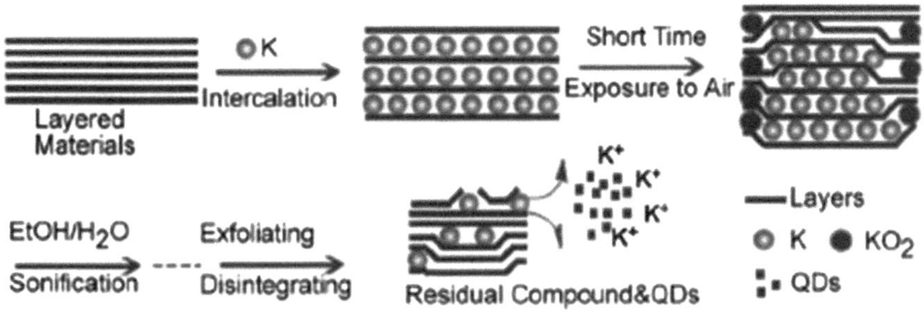

Fig. 2.21 Synthesis of GQDs through potassium-induced exfoliation of carbonaceous layered materials. The scheme shows insertion of the potassium atoms within the carbon layers; subsequent oxidation resulted in fragmentation of the layered materials and formation of GQDs. Reproduced from Lin et al. [23], with permission of the Royal Society of Chemistry

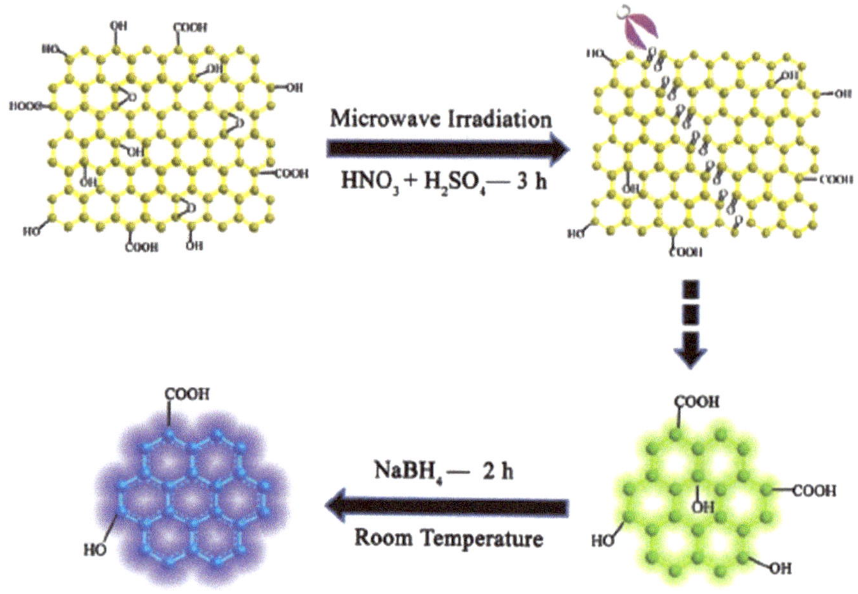

Fig. 2.22 Synthesis of GQDs via combined oxidation in strong acids and reduction. Reprinted from Li et al. [24], with permission from John Wiley & Sons (2012)

route, developed by J-J Zhu and colleagues at Nanjing University, China, is depicted in Fig. 2.22 [24]. The "one-pot" process combined microwave heating of GO and treatment with a strong acid to cleave the GO sheets. Similar to Carbon-Dot preparation schemes, the ubiquitous oxidation step is important for breaking the C–C bonds within the (graphene) starting materials, yielding small nanoparticles exhibiting OH and COOH surface residues. Notably, while acid oxidation produced GQDs exhibiting *green* luminescence, further reaction with sodium borohydride ($NaBH_4$), a commonly used reducing agent, transformed the GQDs' color to *purple*, ascribed to modulation of the surface residues on the nanoparticles.

Other methodologies have been developed for producing GQDs from GO as the starting material. Similar to the above procedure, the main synthetic challenges consist of identifying reaction schemes that are not too harsh to carry out, yielding uniform carbon nanoparticle products in a controlled manner, and utilizing readily-available reagents. Figure 2.23 depicts a facile oxidation/fragmentation method based upon the "Fenton" reaction [25]. The Fenton reaction employs iron ions and hydrogen peroxide to generate hydroxide radicals which constitute powerful oxidizing agents. The hydroxide radicals progressively induced breaking of carbon–oxygen and carbon–carbon bonds in the GO sheets, resulting in formation of GQDs. According to the synthesis model proposed by the researchers and outlined in Fig. 2.23, the Fenton reagent initially induced "holes" in the GO network, which expanded and eventually yielded GQDs. Notably, this pathway might also provide means of controlling the concentrations and sizes of the particles produced.

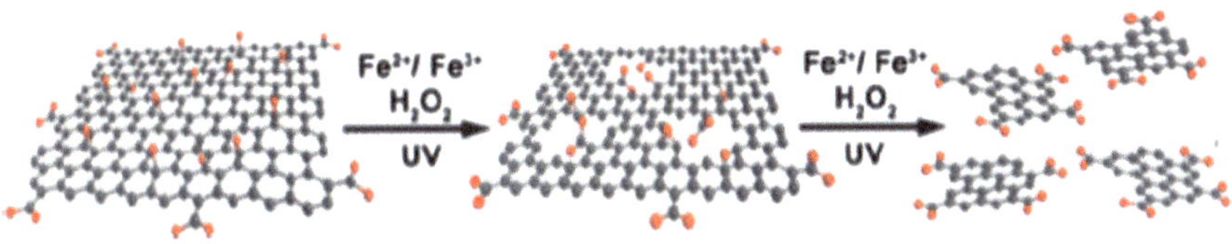

Fig. 2.23 Generation of graphene quantum dots through Fenton reaction of graphene oxide (GO). The scheme shows the fragmentation of the GO sheet following oxidation by the Fenton reagent, ultimately yielding GQDs. Reprinted with permission from Zhou et al. [30]. Copyright (2012) American Chemical Society

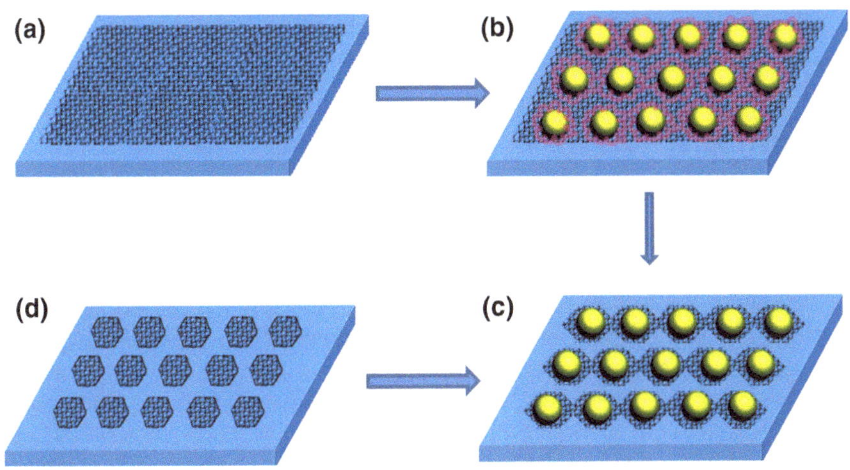

Fig. 2.24 Fabrication of graphene quantum dots using an array of gold nanoparticles and polymer micelles. **a** preparation of a graphene film; **b** an array of polymer micelle-coated Au nanoparticles (NPs) placed upon the film; **c** plasma etching removes the organic substances, leaving behind the Au NPs and graphene fragments underneath; **d** GQDs are produced after removal of the Au NPs. Adapted from Kim et al. [26]

A combined "top-down" and "bottom-up" strategy for production of GQDs is outlined in Fig. 2.24. The methodology, developed by B.H. Sohn and colleagues at Seoul National University, Korea, relied upon placing gold-containing polymer micelles upon a graphene sheet [26]. Following formation of metallic gold nanoparticles within the polymer micelles, the system was subjected to plasma etching (e.g., bombardment with high energy gas molecules). In this scenario, the uniform gold nanoparticles assembled within the polymer "cages" functioned as "shields"—while the plasma treatment removed the unprotected graphene, generating nano-size GQDs underneath the gold particles. Removal of the Au particles subsequently exposed the resultant GQDs. While the approach outlined in Fig. 2.24 involve several steps and appears somewhat elaborate, it enables one to fabricate uniform GQDs in large quantities. Furthermore, control of GQD

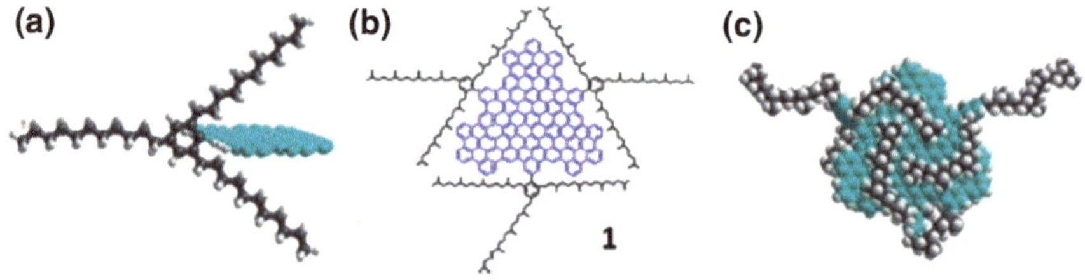

Fig. 2.25 Template-directed synthesis of GQDs. Schemes showing the synthetic concept. **a** A trialkylphenyl residue (*black*) is attached to a graphene sheet (*blue*). Hydrogen atoms involved in stabilization of the GQDs are indicated in *white*; **b** molecular organization of a graphene nanoparticle stabilized by three trialkylphenyl residues; **c** a theoretical calculation depicting the GQD core (*blue*) enclosed by the alkyl chains. Reprinted with permission from Yan et al. [31]. Copyright (2010) American Chemical Society

dimensions is feasible through the choice of polymer "mask" employed, as high-molecular weight polymer substances can generate larger particles and vice versa.

An interesting "templating" synthesis method was developed for production of size-controlled GQDs by L. Li and colleagues at Indiana University (Fig. 2.25). As shown in Fig. 2.25, the researchers "encapsulated" a graphene core within three alkyl-phenyl residues which both determined the dimensionalities of the dots as well as prevented aggregation of GQDs in solution. The resultant GQDs were relatively large (having diameters at around 15 nm), however their uniform sizes gave rise to a defined absorption peak in the visible spectral region, enabling their potential use as light absorbers in solar cells. Moreover, this synthesis strategy makes possible, in principle, tuning of the particles' sizes by varying the *lengths* of the alkyl chains.

References

1. K. Hola, Y. Zhang, Y. Wang, E.P. Giannelis, R. Zboril, A.L. Rogach, Nano Today **9**, 590–603 (2014)
2. J. Zhou, C. Booker, R. Li, X. Zhou, T.-K. Sham, X. Sun, Z. Ding, J. Am. Chem. Soc. **129**, 744–745 (2007)
3. Y.-P. Sun, B. Zhou, Y. Lin, W. Wang, K.A.S. Fernando, P. Pathak, M.J. Meziani, B.A. Harruff, X. Wang, H. Wang, P.G. Luo, H. Yang, M.E. Kose, B. Chen, L.M. Veca, S.-Y. Xie, J. Am. Chem. Soc. **128**, 7756–7757 (2006)
4. L. Cao, X. Wang, M.J. Meziani, F. Lu, H. Wang, P.G. Luo, Y. Lin, B.A. Harruff, L.M. Veca, D. Murray, S.-Y. Xie, Y.-P. Sun, J. Am. Chem. Soc. **129**, 11318–11319 (2007)
5. Y.-P. Sun, X. Wang, F. Lu, L. Cao, M.J. Meziani, P.G. Luo, L. Gu, L.M. Veca, J. Phys. Chem. C **112**, 18295–18298 (2008)
6. H. Yu, X. Li, X. Zeng, Y. Lu, Chem. Commun. **52**, 819–822 (2016)
7. H. Li, X. He, Z. Kang, H. Huang, Y. Liu, J. Liu, S. Lian, C.H.A. Tsang, X. Yang, S.-T. Lee, Water-soluble fluorescent carbon quantum dots and photocatalyst design. Angew. Chem. Int. Ed. **49**, 4430–4434 (2010)

8. J. Lu, J.-X. Yang, J. Wang, A. Lim, S. Wang, K.P. Loh, ACS Nano **3**, 2367–2375 (2009)
9. R. Liu, D. Wu, S. Liu, K. Koynov, W. Knoll, Q. Li, An aqueous route to multicolor photoluminescent carbon dots using silica spheres as carriers. Angew. Chem. Int. Ed. **48**, 4598–4601 (2009)
10. S.C. Ray, A. Saha, N.R. Jana, R. Sarkar, J. Phys. Chem. C **113**, 18546–18551 (2009)
11. L. Tian, D. Ghosh, W. Chen, S. Pradhan, X. Chang, S. Chen, Chem. Mater. **21**, 2803–2809 (2009)
12. Z.-A. Qiao, Y. Wang, Y. Gao, H. Li, T. Dai, Y. Liu, Q. Huo, Chem. Commun. **46**, 8812–8814 (2010)
13. L. Tang, R. Ji, X. Cao, J. Lin, H. Jiang, X. Li, K.S. Teng, C.M. Luk, S. Zeng, J. Hao, S.P. Lau, ACS Nano **6**, 5102–5110 (2012)
14. M. Xu, G. He, Z. Li, F. He, F. Gao, Y. Su, L. Zhang, Z. Yang, Y. Zhang, Nanoscale **6**, 10307–10315 (2014)
15. F. Wang, S. Pang, L. Wang, Q. Li, M. Kreiter, C.-Y. Liu, Chem. Mater. **22**, 4528–4530 (2010)
16. K. Jiang, S. Sun, L. Zhang, Y. Wang, C. Cai, H. Lin, ACS Appl. Mater. Interfaces **7**, 23231–23238 (2015)
17. Y. Hu, J. Yang, J. Tian, L. Jia, J.-S. Yu, Waste frying oil as a precursor for one-step synthesis of sulfur-doped carbon dots with pH-sensitive photoluminescence. Carbon **77**, 775–782 (2014)
18. Z. Zhang, J. Hao, J. Zhang, B. Zhang, J. Tang, RSC Adv. **2**, 8599–8601 (2012)
19. S.K. Bhunia, A.R. Maity, S. Nandi, D. Stepensky, R. Jelinek, ChemBioChem **17**, 614–619 (2016)
20. O. Kozák, K.K.R. Datta, M. Greplová, V. Ranc, J. Kašlík, R. Zbořil, J. Phys. Chem. C **117**, 24991–24996 (2013)
21. H. Hou, C.E. Banks, M. Jing, Y. Zhang, X. Ji, Carbon quantum dots and their derivative 3D porous carbon frameworks for sodium-ion batteries with ultralong cycle life. Adv. Mater. **27**, 7861–7866 (2015)
22. L.A. Ponomarenko, F. Schedin, M.I. Katsnelson, R. Yang, E.W. Hill, K.S. Novoselov, A.K. Geim, Science **320**, 356–358 (2008)
23. L. Lin, S. Zhang, Chem. Commun. **48**, 10177–10179 (2012)
24. L.-L. Li, J. Ji, R. Fei, C.-Z. Wang, Q. Lu, J.-R. Zhang, L.-P. Jiang, J.-J. Zhu, A facile microwave avenue to electrochemiluminescent. Two-color graphene quantum dots. Adv. Funct. Mater. **22**, 2971–2979 (2012)
25. S. Zhuo, M. Shao, S.-T. Lee, ACS Nano **6**, 1059–1064 (2012)
26. K. Sung-Soo, C. Jeong-Yong, K. Kwan, S. Byeong-Hyeok, Nanotechnology **23**, 125301 (2012)
27. X. Xu et al., J. Am. Chem. Soc. **126**, 12736–12737 (2004)
28. K. Jiang et al., Red, green, and blue luminescence by carbon dots: full-color emission tuning and multicolor cellular imaging. Angew. Chem. Int. Ed. **54**, 5360–5363 (2015)
29. R. Liu et al., J. Am. Chem. Soc. **133**, 15221–15223 (2011)
30. X. Zhou et al., ACS Nano **6**, 6592–6599 (2012)
31. X. Yan et al., Nano Lett. **10**, 1869–1873 (2010)

Chapter 3
Characterization and Physical Properties of Carbon-Dots

As new members of the "nanoparticle universe," Carbon-Dots have been extensively characterized using varied analytical methods. Moreover, since Carbon-Dots exhibit interesting optical properties, numerous studies have aimed to elucidate the mechanisms responsible for the photophysical phenomena associated with the carbon nanoparticles. Below I only briefly summarize experimental techniques applied to characterize Carbon-Dots and the structural/chemical information generated by the measurements. A detailed discussion is also presented focusing on physical features of Carbon-Dots, particularly the intriguing *luminescence* phenomena, and representative studies designed to probe the underlying factors responsible for these properties. *Physical processes* aimed at modulating Carbon-Dots' luminescence are additionally outlined.

3.1 Structural Characterization

Varied analytical methods are routinely applied to characterize Carbon-Dots and their physical properties, to illuminate the crystalline organization of the carbon atoms, and to probe the type and abundance of functional units displayed upon the Carbon-Dots' surface. Characterization of Carbon-Dots is essential for attaining better understanding of the mechanisms associated with the unique physical properties of these nanoparticles.

Transmission electron microscopy (***TEM***) has been a primary technique for visualization of Carbon-Dots, providing important information upon particle morphology, size distribution, and crystalline organization. ***High-resolution TEM*** (***HRTEM***) experiments have been applied to confirm the *periodicity* of the graphitic core, reflecting its crystalline nature. Figure 3.1 presents a HRTEM image of a Carbon-Dots, featuring the graphite lattice planes of the carbon core. ***X-ray diffraction*** (***XRD***) is another important structural tool for evaluation of the crystalline nature of Carbon-Dots. XRD provides information upon the unit cell

© Springer International Publishing Switzerland 2017

R. Jelinek, *Carbon Quantum Dots*, Carbon Nanostructures,
DOI 10.1007/978-3-319-43911-2_3

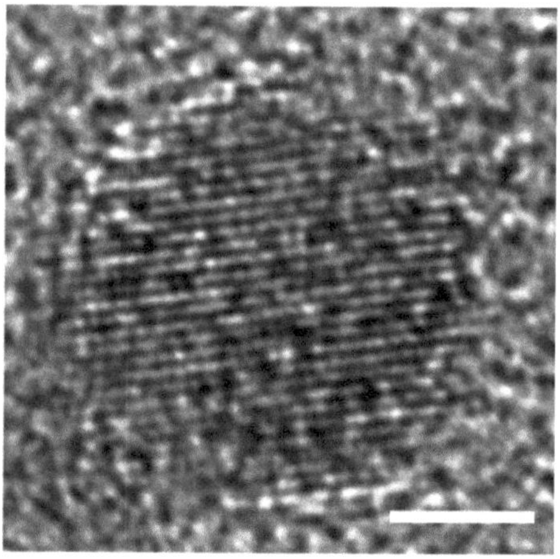

Fig. 3.1 High-resolution transmission electron microscopy (HRTEM) of a carbon-dot. The crystal planes of the graphitic carbon core are clearly visible. *Scale bar* corresponds to 2 nm

dimensions and crystal spacing within the crystalline carbon cores. Figure 3.2 depicts a representative XRD pattern, showing a diffraction peak at around 20° reflecting the crystalline graphitic structure. The XRD peak in Fig. 3.2 points, in fact, to a lattice spacing of 0.45 nm which is *greater* than bulk graphite (~0.35 nm)—indicating certain amorphous character of the synthesized Carbon-Dots.

Raman scattering reveals the structural features of the carbon atoms within Carbon-Dots. A typical Raman spectrum of Carbon-Dots is shown in Fig. 3.3 featuring two peaks corresponding to the D and G bands, respectively. The D band at around 1350 cm^{-1} is ascribed to *disordered* sp^2 *carbons*, while the G band at around 1600 cm^{-1} arises from the in-plane stretching vibration mode E_{2g} of *crystalline graphite carbons*. The ratio of the intensities (I_D/I_G) of the characteristic Raman bands can be used to study the structural properties of the carbon framework, particularly the degree of crystallinity and relative abundance of *core* carbon atoms versus *surface* atoms.

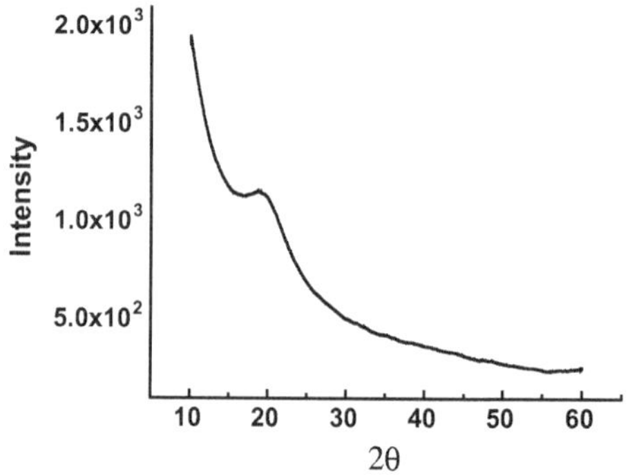

Fig. 3.2 X-ray diffraction pattern of a Carbon-Dots sample. The weak broad reflection peak centered at around 19° corresponds to an interlayer spacing of approximately 0.45 nm and reflects relatively poor crystallinity

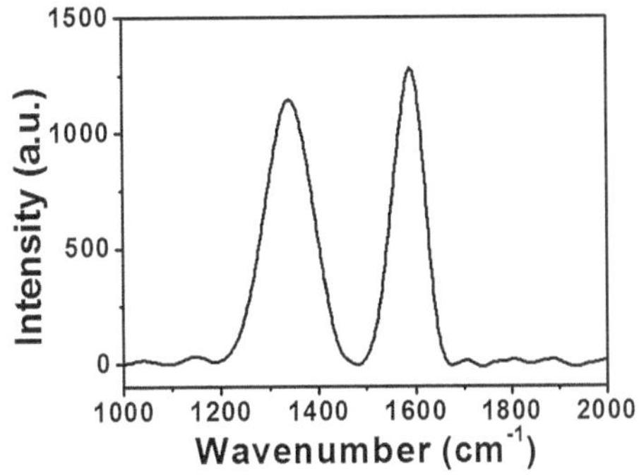

Fig. 3.3 Raman spectrum of Carbon-Dots sample. Shown are the D and G bands at ~1340 and 1590 cm^{-1}, respectively. The I_G/I_D ratio in this case is 1.12, indicating more abundant graphitic carbon atoms

Elucidating the functional groups upon Carbon-Dots' surfaces is carried out through application of several widely used analytical methods. ***X-ray photoelectron spectroscopy*** (***XPS***) provides information upon specific atomic units present upon Carbon-Dots' surface. An example of an XPS analysis is provided in Fig. 3.4. The spectral analysis reveals the distinct nitrogen-, oxygen-, and carbon-bonded units displayed upon the Carbon-Dots's surface. ***Fourier transform infrared*** (***FTIR***) ***spectroscopy*** usually complements XPS, illuminating distinct functional units through recording of typical vibration bands (Fig. 3.5).

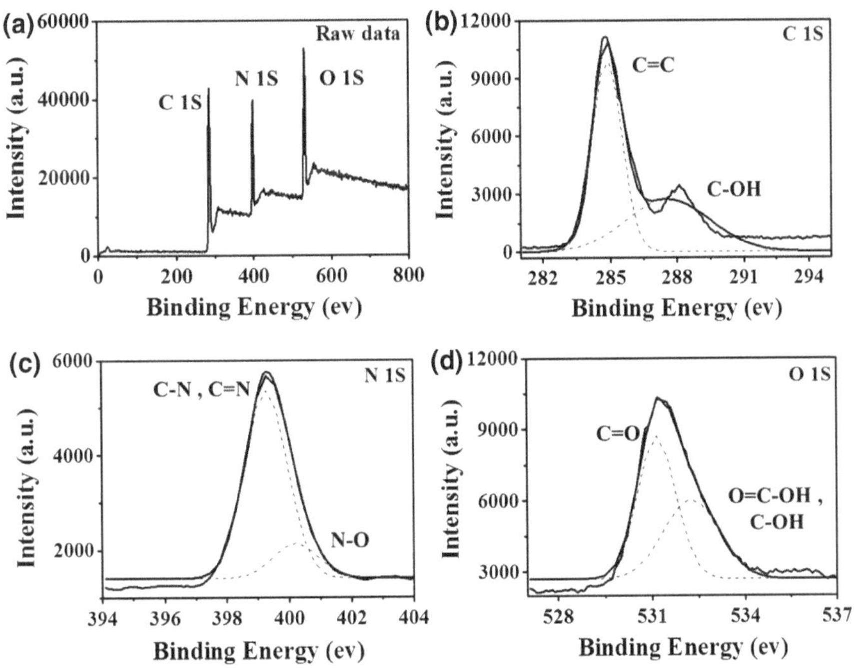

Fig. 3.4 X-ray photoelectron spectra (XPS) of carbon-dots. The spectra reveal distinct functional groups corresponding to different atoms present upon the Carbon-Dots' surface

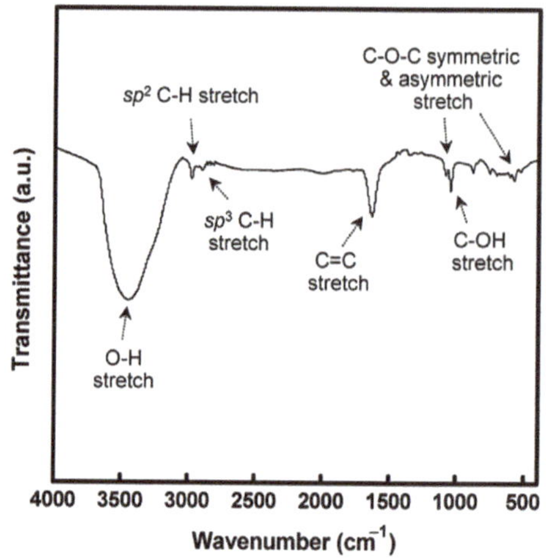

Fig. 3.5 Fourier-transform infrared (FTIR) spectrum of a carbon-dot sample. Distinct vibration bands corresponding to Carbon-Dots' surface units are indicated

3.2 Photophysical Properties

Perhaps the most striking property of Carbon-Dots is the *excitation-dependent photoluminescence* (*luminescence*), often referred to as excitation-dependent fluorescence emission. Typical excitation-dependent luminescence spectra and the corresponding colors are depicted in Fig. 3.6. The wide spectral range and relatively high intensities of the emission peaks underline the *multicolor* properties of Carbon-Dots. Indeed, the fact that the emission color of Carbon-Dots can be tuned according to the excitation wavelength is among the unique properties of Carbon-Dots, exploited in varied applications.

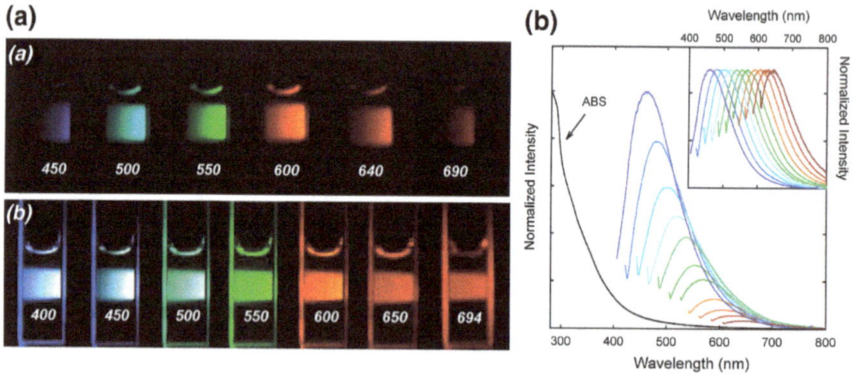

Fig. 3.6 Excitation-dependent luminescence of carbon-dots. **a** *Photographs* of polyethylene glycol (PEG)-passivated Carbon-Dots excited at 400 nm (*a*), and recorded in different bandpass emission filters, or excited at the indicated wavelengths (*b*). **b** Fluorescence emission spectra of the Carbon-Dots excited at 20-nm increasing increments. The inset shows the emission spectra at normalized intensities. Reprinted with permission from Sun et al. [13]. Copyright (2006) American Chemical Society

The remarkable phenomenon shown in Fig. 3.6 has attracted considerable interest both from fundamental physical and practical applications standpoints. While the widely studied emission colors of *semiconductor* quantum dots (Qdots)—the hallmarks of these nanoparticles—correspond to the quantum confinement effects and correlate to the nanoparticles' *sizes* (i.e., diameters), the underlying properties responsible for the luminescence properties of Carbon-Dots is different. While size variations are encountered in Carbon-Dots synthesis schemes (as well as graphene quantum dots), the origin of the distinct colors observed in Carbon-Dots has been ascribed, in fact, to ***energy states*** associated with ***surface defects*** upon the graphitic nanoparticles. Figure 3.7 illustrates this concept, showing fluorescence emissions generated from distinct energy levels associated with Carbon-Dots' surface.

Many studies have provided experimental evidence supporting the luminescence model illustrated in Fig. 3.7 and showing that modulation of the "surface states" of Carbon-Dots is the underlying source for the excitation-dependent luminescence properties. Indeed, as discussed above in Chap. 2 (Carbon-Dots Synthesis), early research in the Carbon-Dots field has shown that luminescence could be achieved *only* after surface functionalization of the carbon nanoparticles through post-synthesis chemical reactions (also termed "surface passivation"). In fact, several studies have shown that Carbon-Dots exhibiting *single* emission wavelengths (e.g., no excitation-dependent luminescence) could be synthesized through careful modification of the Carbon-Dots' surface, thereby eliminating or reducing the concentrations of the "surface defects" responsible for the different energy levels.

Modulation of the structural/functional properties of Carbon-Dots through chemical means has been a prominent research direction in the field, aimed at controlling the excitation-dependent emission profiles of Carbon-Dots. H. Lin and colleagues at the Ningbo Institute of Materials Technology and Engineering, China, for example, synthesized Carbon-Dots for which the emission spectra exhibited *similar peak intensities* even upon excitation in significantly different wavelengths [1] (Fig. 3.8). Such "full color" Carbon-Dots are, in fact, different than the overwhelming majority of Carbon-Dots systems reported in the literature—in which high emission peak intensities have been generally observed only within a specific, quite narrow fluorescence "window" (and rapid decrease in fluorescence emission

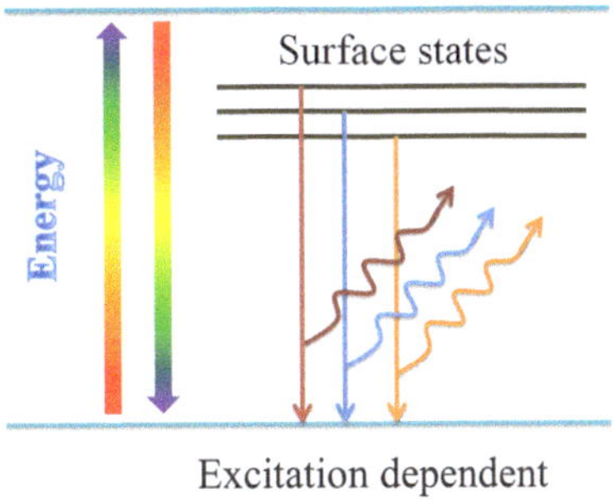

Fig. 3.7 Variable-color photoluminescence of Carbon-Dots ascribed to different surface energy states

signals upon excitation outside of the spectral window, see for example Fig. 3.6b). Important from a mechanistic standpoint, the multicolor properties of the Carbon-Dots presented in Fig. 3.8 were traced to the relative abundance of functional groups upon the Carbon-Dots' surface, particularly *aromatic* residues, *C=N*, *C=O*, and *C–N* bonds. Indeed, the authors demonstrated that changing the formamide-generated synthesis conditions yielded Carbon-Dots exhibiting different distributions of the chemical units upon the surface, and corresponding changes in the emission colors.

The close relationship between Carbon-Dots' luminescence and surface functional groups has been manifested in many other studies. An intriguing example of fluorescence "on-off" in Carbon-Dots, accomplished through *esterification* reactions of surface residues, is outlined in Fig. 3.9. The Carbon-Dots, prepared by D. Chowdhury and colleagues at the Institute of Advanced Study in Science and Technology, India, from citric acid as a carbon precursor, exhibited diminished

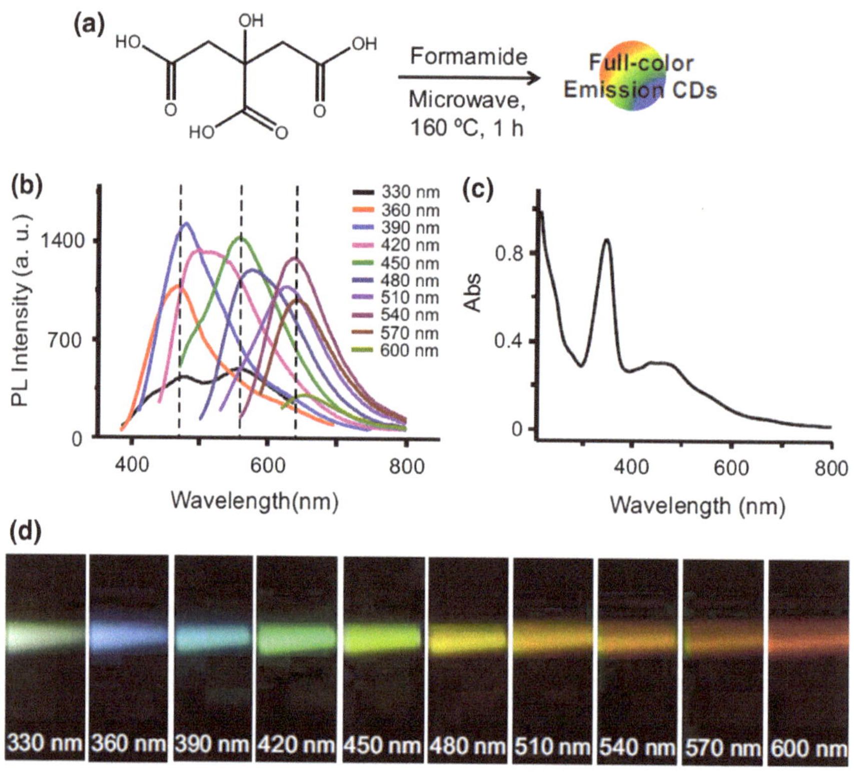

Fig. 3.8 "Full-color" Carbon-Dots. **a** Synthesis scheme; **b** fluorescence emission peaks upon different excitation wavelengths. Note the similar peak intensities. **c** UV–vis absorbance spectrum showing a broad absorbance peak ascribed to the presence of multiple energy absorbing residues upon the Carbon-Dots' surface. **d** Luminescence colors reflecting the excitation-dependent emission spectra. Reprinted from Pan et al. [1], with permission from John Wiley & Sons (2015)

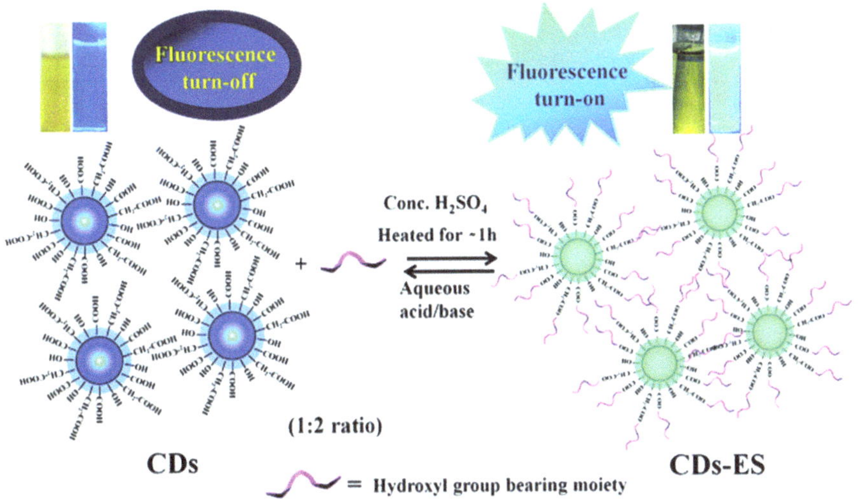

Fig. 3.9 Modulation of Carbon-Dots' fluorescence induced by esterification of surface residues. The fluorescence of the Carbon-Dots was low in citric acid-generated Carbon-Dots at pH = 1 (*left*); upon esterification with alcohol-bearing moieties, the fluorescence was significantly increased (*right*). Reproduced from Baruah et al. [2], with permission of the Royal Society of Chemistry

fluorescence in very acidic conditions [2]. However, upon *esterification* of the hydroxide and carboxylic acid residues through reaction with alcohols, the Carbon-Dots became highly fluorescent. Interestingly, the process was *reversible*—acidification of the esterified Carbon-Dots resulted in fluorescence quenching, subsequent re-esterification again recovered the fluorescence emission, etc. While the mechanism responsible for this phenomenon has not been clarified, the fluorescence intensity modulation has been ascribed to changes in the surface energy states, directly affected by the chemical reactions.

Many studies have proposed mechanistic descriptions of the excitation-dependent luminescence of Carbon-Dots. A.S. Urban and colleagues at the Ludwig-Maximilians-Universitaet, Germany, conducted a careful comparison between the optical properties of Carbon-Dots embedded within a polymer matrix and poly-aromatic hydrocarbons (PAHs) that were similarly encapsulated within the polymer host, enabling control over concentrations and compositions of the emitting species [3]. The model proposed by the researchers, outlined in Fig. 3.10, treats the Carbon-Dots as a "cocktail" of two-dimensional PAH fragments (e.g., molecular units containing carbon in the sp^2 hybridization) embedded within a matrix of amorphous carbon. The luminescence in this configuration was due to excitation of the aromatic species, and the different emission colors arose from slight structural modifications of the PAH units, particularly carbon functionalization with oxygen and nitrogen. The structural differences among the PAHs resulted in distinct energy gaps, corresponding to the different emission wavelengths.

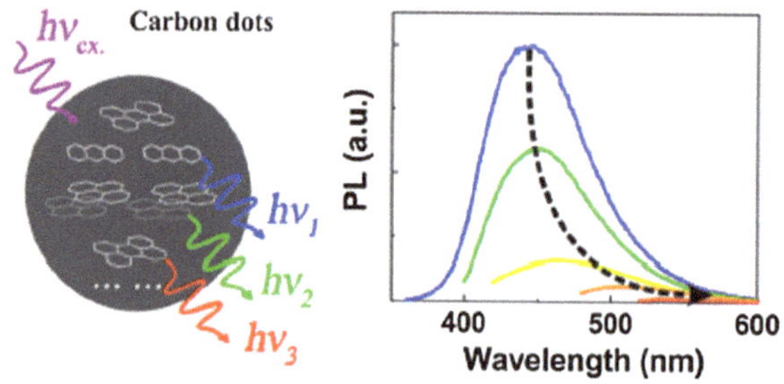

Fig. 3.10 Mechanistic model of the multicolor properties of Carbon-Dots comprising poly-aromatic hydrocarbons (PAHs). The scheme on the *left* shows excitation of the Carbon-Dots at a specific wavelength and emission in several colors (wavelengths) due to variations in PAH structures. The spectra on the *right* reflect the different luminescence emissions. Reprinted with permission from Fu et al. [3]. Copyright (2015) American Chemical Society

Energy transfer among the PAHs might further contribute to variations in luminescence within the Carbon-Dots system.

In light of the centrality of the Carbon-Dots' surfaces to their photophysical properties, varied synthetic modifications designed to probe the specific contributions of "surface defects" have been pursued. G. Shi and colleagues at Tsinghua University, China, demonstrated *tuning* of the excitation/emission properties of graphene quantum dots (QQDs) through reaction with *aryl* residues (e.g., an aromatic ring covalently linked to different functional groups) [4]. Based upon the experimental data, the researchers attempted to pinpoint the surface sites responsible for the luminescence properties of the GQDs, and further proposed a mechanism pertaining to the observed shifts in fluorescence emissions and intensity modifications induced by the aryl moieties (Fig. 3.11). Specifically, according to the proposed model outlined in Fig. 3.11b, the excitation and emission in the GQD system were related to the HOMO–LUMO transition associated with the two free electrons of the sp^2-hybridized *carbene* units upon the GQDs' surface. Reaction of the carbenes with aryl groups yielded *radicals* for which the energy transition was significantly modulated (corresponding to *shifts* in fluorescence emission). Furthermore, coupling of the aryls reduced the concentration of *paired electrons* which were responsible for quenching of the fluorescence emission, thus leading to an overall higher fluorescence intensity. Both scenarios were consistent with the observed luminescence modulation and point to the central roles played by the *surface electrons* of Carbon-Dots in determining the optical/spectroscopic properties of the nanoparticles.

Other studies aimed to delineate the mechanisms responsible for generation of surface energy states in Carbon-Dots. A key in many such studies has been to effectively isolate Carbon-Dots exhibiting defined chemical and photophysical features. H.M. Xiong and colleagues in Fudan University, China, prepared Carbon-Dots exhibiting fluorescence emissions covering the entire visible spectral range [5] (Fig. 3.12). Crucially, those Carbon-Dots could be separated to *optically pure*

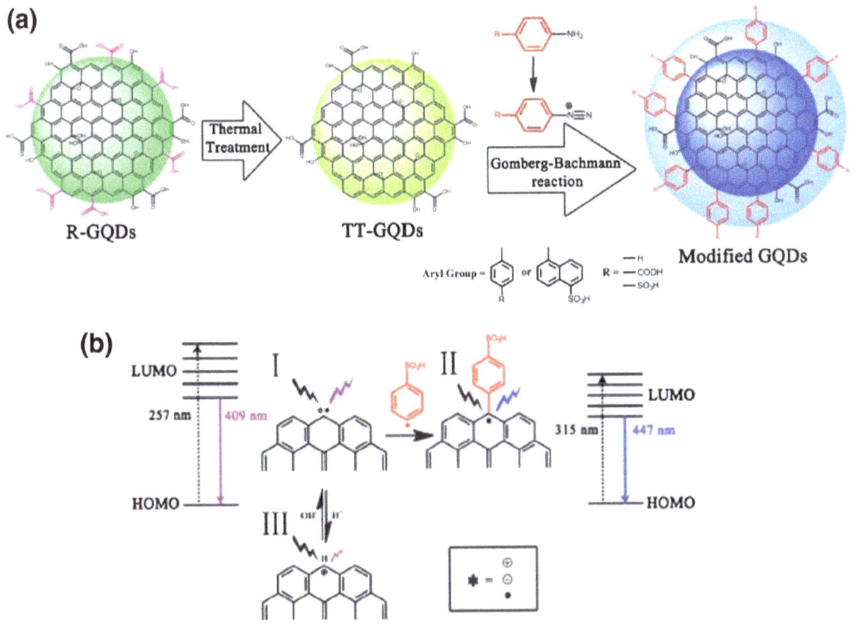

Fig. 3.11 Luminescence modulation in aryl-functionalized graphene quantum dots (GQDs). **a** Reaction scheme. The aryl residues are covalently linked to surface carbon atoms. **b** Model depicting the fluorescence modulation upon aryl reaction with carbene units. The chemical reaction results both in shift of the fluorescence emission due to difference in the HUMO–LUMO energy spacing, as well as greater emission intensity associated with induction of radical recombination through electron pairing. Reproduced from Luo et al. [4], with permission of the Royal Society of Chemistry

particles emitting distinct luminescent colors using silica column chromatography, because each Carbon-Dots species exhibited distinct polarity. The isolated Carbon-Dots were analyzed and their fluorescence emissions were found to be directly correlated with the degree of *surface oxidation* of the dots, rather than parameters such as the Carbon-Dots size (thus excluding possible "quantum size" effects similar to inorganic semiconductor nanoparticles). Figure 3.12d depicts the proposed relationship between the Carbon-Dots' colors and degree of surface oxidation.

While the publications discussed above suggest that surface states and/or "surface defects" are the predominant physical features affecting the excitation-dependent emission of Carbon-Dots, other theories have been put forward as well. J. Zhang and colleagues at East China University of Science and Technology, for example, size-separated a GQD sample using gel electrophoresis, accomplishing well-defined and narrow size distributions of the particles [6]. Examination of the fluorescence properties of the different Carbon-Dots fractions revealed that *excitation-dependent emission* was apparent only in case of the *smaller* nanoparticles, <5 nm (Fig. 3.13a). In comparison, *larger* GQDs exhibited

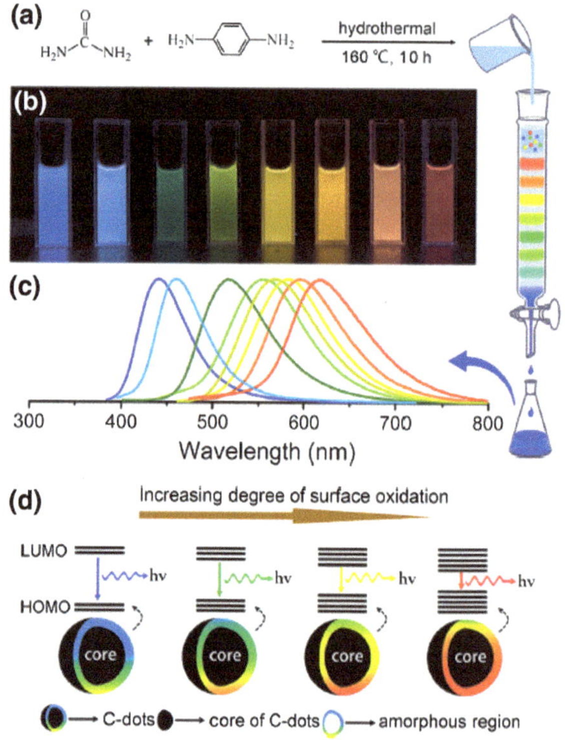

Fig. 3.12 Carbon-dots exhibiting specific luminescence emission wavelengths. **a** synthesis using phenylenediamine precursor and purification of Carbon-Dots exhibiting distinct colors by column chromatography; **b** colors of different Carbon-Dots fractions; **c** emission wavelengths; **d** hypothesized mechanism linking the emission colors to degree of surface oxidation. Reprinted with permission from Ding et al. [15]. Copyright (2016) American Chemical Society

the *same* emission upon different excitations (Fig. 3.13b). The researchers hypothesized that in the small GQDs the fluorescence emission was ascribed to surface energy states, as the surface area is predominant compared to the crystalline core. Accordingly, even slight variations of the abundance of functional units such as carboxylic acids would give rise to different excitation/emission spectra, as apparent in Fig. 3.13a. In the bigger GQDs, however, the emission is largely affected by the carbon scaffold; accordingly, the narrow size distribution of the particles yielded excitation-*independent* emission spectra. Overall, this study suggests that minute variations in Carbon-Dots *sizes* might also significantly contribute to the excitation-dependent emission phenomena.

Fractionation of Carbon-Dots samples was used in other studies to obtain Carbon-Dots that exhibited distinct luminescence properties, shedding light on the parameters and mechanisms responsible for their excitation-dependent emission. L. A. Colon and colleagues at SUNY, Buffalo, for example, revealed that fractions collected upon passing an as-synthesized Carbon-Dots sample through a high-performance liquid chromatography (HPLC) column displayed *different* fluorescence profiles compared to the "parent" Carbon-Dots [7]. Specifically, the Carbon-Dots fractions featured sharper emission spectra and greater quantum yields and brightness (Fig. 3.14). Particularly important, the researchers reported that the Carbon-Dots fractions produced excitation-*independent* fluorescence emission (in

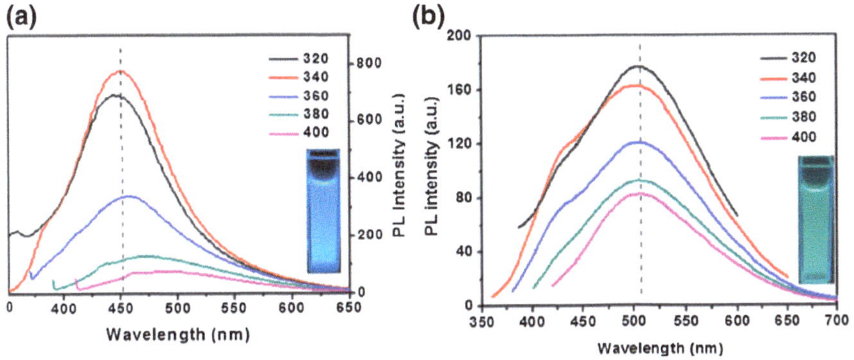

Fig. 3.13 Size dependence of the excitation/emission properties of graphene quantum dots. **a** GQDs exhibiting small, <5 nm, sizes—excitation-dependent emission is observed; **b** Larger GQDs give rise to excitation-*independent*, constant emission wavelength. The insets correspond to the luminescence upon UV irradiation (302 nm). Reprinted with permission from Zhang et al. [6]. Copyright (2016) American Chemical Society

contrast to the excitation-dependent emission spectra generated by the *as-synthesized* dots). This dramatic result is consistent with the notion that the excitation-dependent emission phenomena arise from the presence of many fluorophores (each exhibiting distinct structural features) in a sample.

The relationship between Carbon-Dots' luminescence and their surface properties facilitates interesting applications, such as distinguishing between different electron *acceptor* molecules in solution. Figure 3.15 depicts experimental data reported by Y.P. Sun and colleagues at Clemson University, US, demonstrating *quenching* of Carbon-Dots' fluorescence by 4-nitrotoluene and 2,4-dinitrotoluene, respectively. Notably, the degrees of quenching the *fluorescence intensity* and *fluorescence lifetime* were both different between the two compounds, reflecting their different electron accepting properties. Specifically, the researchers hypothesized that the distinct quenching profiles might be ascribed to disruption of electron–hole radiative recombination processes associated with surface residues that are responsible for the fluorescence emissions.

3.3 Physical Processes Associated with Carbon-Dots' Luminescence Properties

Varied phenomena involving Carbon-Dots' luminescence have been investigated. ***Electrochemiluminescence*** (***ECL***), for example, has been recorded in Carbon-Dots systems. ECL processes occur when electrochemically generated electrons and holes give rise to excited states in the target molecules (Carbon-Dots in the case here) which subsequently relax to their ground states upon emission of light. ECL

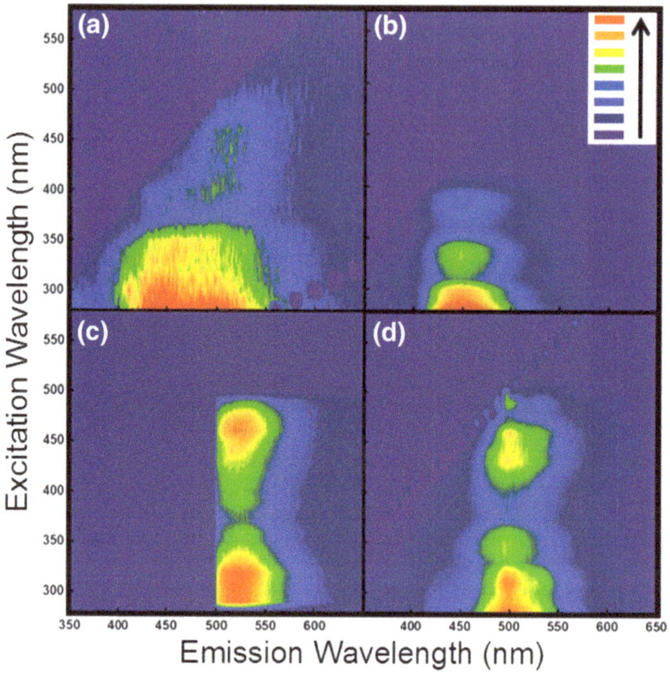

Fig. 3.14 Fractionation of a Carbon-Dots sample alters the luminescence properties. Excitation/emission "topographic maps" recorded for the as-synthesized Carbon-Dots sample (**a**), and three different fractions (**b–d**). Note that the emission profiles of all Carbon-Dots fractions were significantly *narrower* compared to the parent, as-synthesized sample. Reprinted with permission from Vinci et al. [7]. Copyright (2013) American Chemical Society

of Carbon-Dots was reported by Y. Chi and colleagues at Fuzhou University, China, through electrochemical processing of graphite rods [8]. The proposed ECL mechanism in the Carbon-Dots system is outlined schematically in Fig. 3.16 and largely mimics known ECL mechanisms in semiconductor nanoparticles. In principle, *surface states* of the Carbon-Dots were believed to play a critical role in generation of the excited electronic state, subsequently relaxing to a lower surface energy state through light emission. Interestingly, the *shorter wavelength* luminescence observed by the researchers in the same Carbon-Dot samples was ascribed to presumably greater energy separation of *core energy levels* of the Carbon-Dots (Fig. 3.16). This interpretation, however, is controversial since, as discussed in length above, Carbon-Dots luminescence has been ascribed mostly to *surface states* rather than the carbon cores.

While ECL was observed in other Carbon-Dots systems, the exact mechanisms responsible for electrochemically induced luminescence are still debated. J.J. Zhu and colleagues at Nanjing University, China, have demonstrated ECL in a solution of GQDs produced through microwave/acid treatment [9] (Fig. 3.17). However, the researchers reported that ECL could be generated only in the presence of potassium

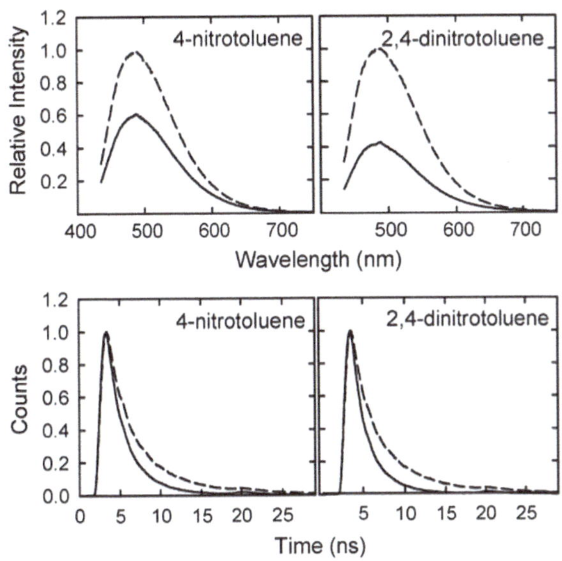

Fig. 3.15 Quenching of Carbon-Dots' fluorescence by electron acceptors. Distinct quenching rates of the fluorescence intensity (*top panels*) and fluorescence lifetimes (*bottom panels*) by two electron acceptors. The results likely reflect the different electron accepting properties of the two compounds. Reproduced from Wang et al. [14], with permission of the Royal Society of Chemistry

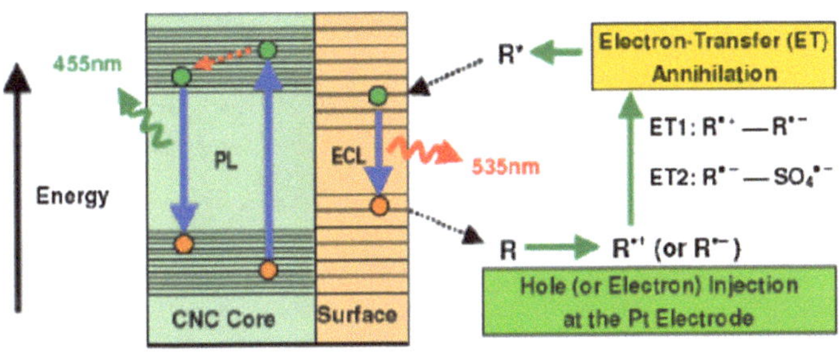

Fig. 3.16 Electrochemiluminescence (ECL) of Carbon-Dots. Proposed model for ECL in Carbon-Dots generated upon electrochemical processing of graphite rods. According to the model light is emitted upon generating excited energy states both in the *core* of the Carbon-Dots (termed "carbon nanocrystal", CNC), as well as upon the *surface*. Reprinted with permission from Zheng et al. [8]. Copyright (2009) American Chemical Society

persulfate ($K_2S_2O_8$) pointing to participation of the persulfate ions in generation of the GQDs' excited state. Figure 3.17 outlines the proposed ECL mechanism in the GQD solution. Specifically, it was hypothesized that a reaction between electrochemically induced *radicals*—GQD$^{\cdot}$ and SO_4^{-}—was in fact responsible for the excited GQD* particles which subsequently emitted luminescence.

ECL was recently reported in Carbon-Dots samples co-dissolved with surfactant molecules in organic solvents (Fig. 3.18). Interestingly, T. Wang and colleagues at the Chinese Academy of Science observed that the Carbon-Dots' ECL could be modulated, depending upon the electrolyte surfactants co-incubated with

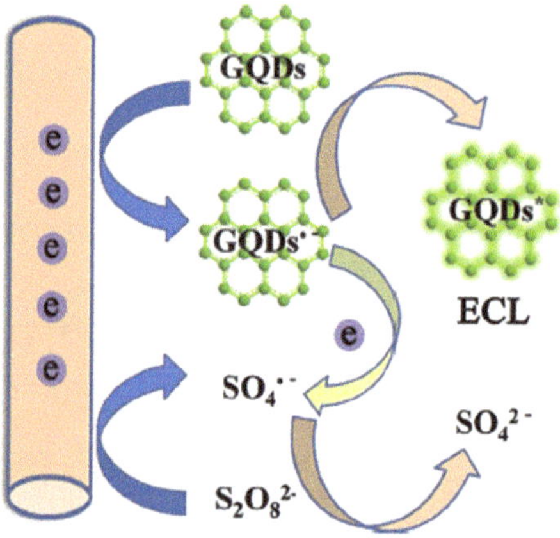

Fig. 3.17 Electrochemiluminescence of graphene quantum dots induced through a radical reaction mechanism. ECL presumably occurs through electron transfer from GQD radical to sulfate radical. Reprinted from Li et al. [9], with permission from John Wiley & Sons (2012)

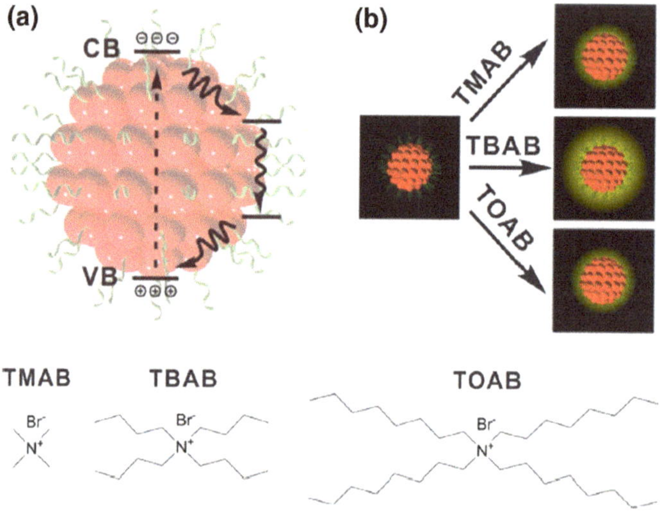

Fig. 3.18 Electrochemiluminescence incurred through interactions of Carbon-Dots with surfactants. **a** Mechanism of electron–hole recombination through surface interactions of the Carbon-Dots with co-dissolved surfactants; **b** scheme representing the different luminescence generated upon reactions of the Carbon-Dots with surfactants (their molecular structures are indicated below). Reprinted with permission from Zhang et al. [10]. Copyright (2014) American Chemical Society

the nanoparticles [10]. According to the proposed mechanism for this phenomenon, the electrochemically-excited electrons within the Carbon-Dots were annihilated through combination with holes formed through reaction with the surfactant electrolytes (Fig. 3.18). The ECL peak was additionally sensitive to concentrations of *metal ions* in the solution, particularly Fe^{3+}, making the system an effective ion sensor.

Chemical luminescence (***CL***) is a phenomenon related to ECL, albeit less investigated in Carbon-Dots systems. CL involves excitation of the Carbon-Dots (i.e., generation of electrons and holes) through means of *chemical reactions*; accordingly, the degree of CL depends both upon the occurrence and yields of the redox reactions responsible for the electron/hole transfer, as well as the efficiency of *coupling* between these reactions and the Carbon-Dots. A generic Carbon-Dots CL mechanism is depicted in Fig. 3.19. In principle, two chemical reactions in solution need to occur in parallel, one responsible for transfer of *electrons* to an excited energy level of the Carbon-Dots (termed "electron injection") while the other reaction *scavenges* an electron from a lower energy level of the Carbon-Dots (i.e., "hole injection"). A consequent *luminescence* occurs upon transition of the injected electrons from the excited energy level to a lower level.

CL in Carbon-Dots systems has been employed for detection of varied analytes through their participation in oxidation/reduction reactions which contribute the electrons and holes. A surprising CL-associated luminescence phenomenon has been recently reported in which high concentrations of *base* molecules presumably led to short wavelength emission in Carbon-Dots (Fig. 3.20). In that study, L. Zhao and colleagues at the Chinese Academy of Science reported that the Carbon-Dots emitted significant high energy luminescence in strongly alkaline solutions [11]. The researcher proposed a mechanism in which the base molecules (such as NaOH) induced "chemical reduction" of residues at the Carbon-Dots' surface, resulting in electron injection into putative surface "electron traps." In parallel, holes were thermally generated upon the Carbon-Dots' surface and the combined effect of electron injection and hole formation resulted in CL. While CL processes help to shed light upon Carbon-Dots' photophysical properties, their practical applicability as a sensing tool has been limited due to the often complex redox reactions and their coupling to the Carbon-Dots, and the practical constrains of the technique in term of reaction yields and luminescence intensities.

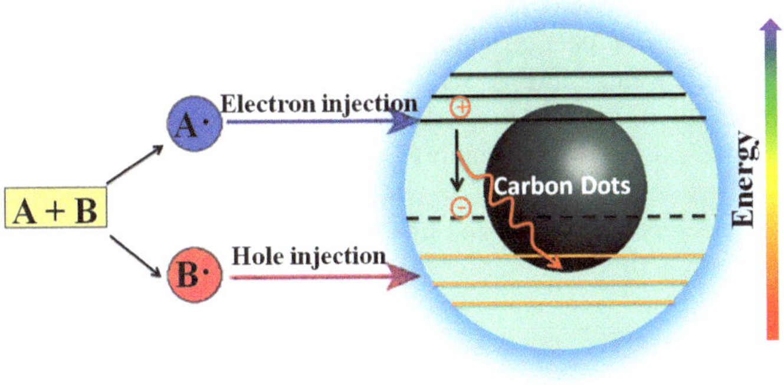

Fig. 3.19 Chemiluminescence (CL) in carbon-dots. Radicals are scavenged by the Carbon-Dots, generating electrons and holes which subsequently emit light upon recombination

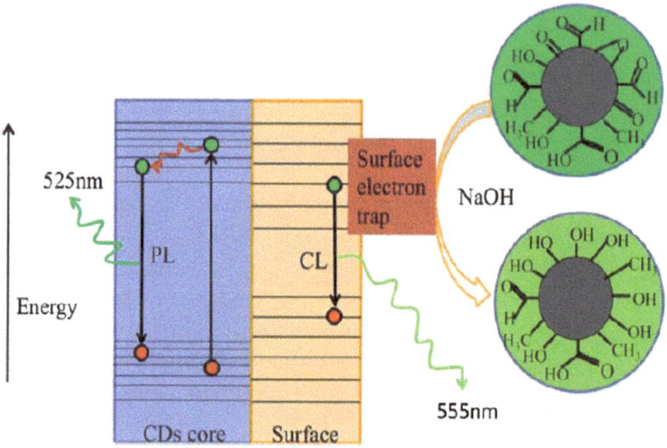

Fig. 3.20 Chemiluminescence of Carbon-Dots in strongly alkaline solutions. Proposed mechanism of the CL associated with surface states of the Carbon-Dots generated upon reaction with NaOH. A photoluminescence (PL) generated from the carbon cores of the Carbon-Dots is also illustrated. Reproduced from Zhao et al. [11], with permission of the Royal Society of Chemistry

Forster resonance energy transfer (***FRET***) is a widely studied fluorescence phenomenon that has been also demonstrated in Carbon-Dots systems. FRET constitutes a non-radiative energy transfer between two fluorophores—fluorescence *donor* and *acceptor*—that are physically adjacent to each other. Thus, the technique has been useful for assessing structural features of materials, particularly biological molecules. Common examples of FRET applications have been in monitoring modulation of *intramolecular* distances occurring through protein *folding* or *misfolding* processes. FRET has been also employed to monitor *membrane reorganization* through embedding the donor and acceptor molecules within the lipid bilayer framework.

Carbon-Dots, in fact, might constitute excellent candidates for functioning as both *donors* and *acceptors* in FRET experiments, as they exhibit a broad range of excitation/emission wavelengths. This property implies that a *single* Carbon-Dots species can serve as an energy donor to *different* acceptors—simply using a different excitation wavelength (Fig. 3.21). Similarly, Carbon-Dots might constitute a FRET acceptor of different *donors*. Figure 3.21b illustrates an example of this concept in a *lipid bilayer* system comprising hydrophobic Carbon-Dots which serve as energy donors to two *different* acceptor molecules, each excited at a distinct wavelength. Specifically, in the experiments outlined in Fig. 3.21b efficient FRET was recorded in both experiments, manifested in the increase of fluorescence emissions of the two acceptor dyes, respectively, upon excitation of the Carbon-Dots [12]. The data in Fig. 3.21 point to possible use of Carbon-Dots-facilitated FRET for analyses of membrane dynamics and structural transformations induced, for example, by membrane-active molecules.

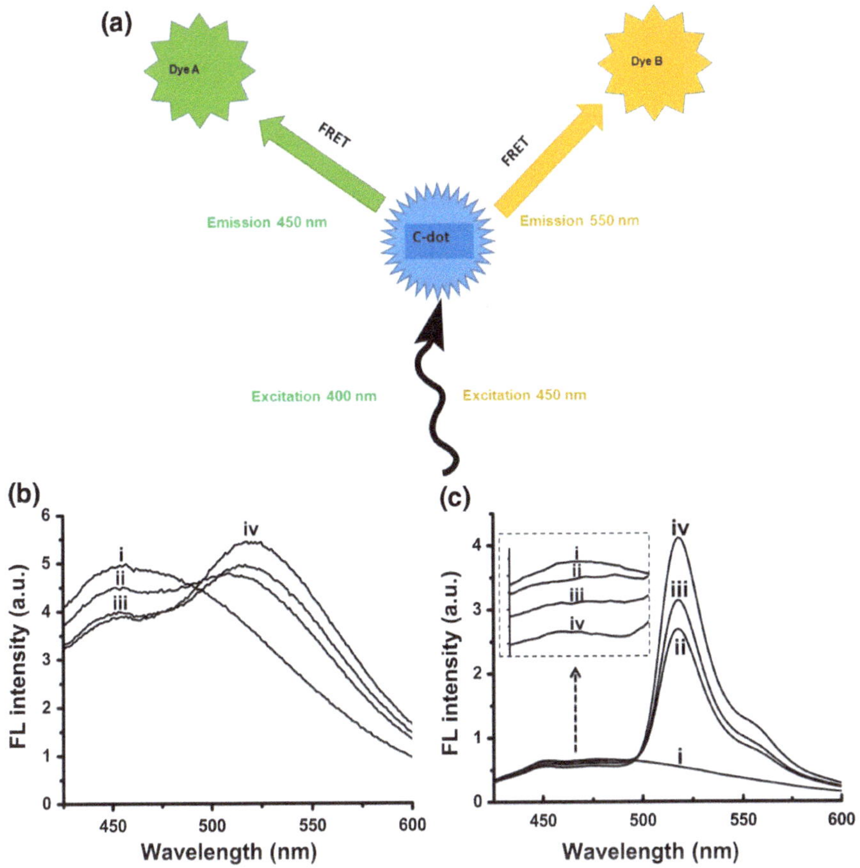

Fig. 3.21 Forster resonance energy transfer (FRET) employing carbon-dots as energy donors. **a** General scheme of FRET experiments employing Carbon-Dots as energy donors to acceptor dyes exhibiting different excitation wavelengths. The excitation/emission wavelengths indicated are arbitrary. Experimental results showing FRET recorded in lipid vesicles containing hydrophobic Carbon-Dots as energy donors and two *different* membrane-associated dyes as energy acceptors: **b** *N*-(7-nitrobenz-2-oxa-1,3-diazol-4-yl)1,2-dihexadecanoyl-sn-glycero-3-phosphoethanolamine (NBD-PE; excitation maximum 469 nm, emission maximum 540 nm), **c** 4,4-Difluoro-8-(2-(2-((1,3-dioxoisoindolin-2-yl)oxy)acetamido)phenyl)-1,3,5,7-tetramethyl-4-bora-3a,4a-diaza-s-indacene-Phthalamide (BODIPY-PH; excitation maximum 500 nm, emission maximum 510 nm). Both in **(b)** and **(c)**, higher concentrations of the Carbon-Dots (i - iv) result in greater fluorescence emission intensity from the acceptor dye

References

1. L. Pan, S. Sun, A. Zhang, K. Jiang, L. Zhang, C. Dong, Q. Huang, A. Wu, H. Lin, Truly fluorescent excitation-dependent carbon-dots and their applications in multicolor cellular imaging and multidimensional sensing. Adv. Mater. **27**, 7782–7787 (2015)
2. U. Baruah, M.J. Deka, D. Chowdhury, RSC Adv. **4**, 36917–36922 (2014)

3. M. Fu, F. Ehrat, Y. Wang, K.Z. Milowska, C. Reckmeier, A.L. Rogach, J.K. Stolarczyk, A.S. Urban, J. Feldmann, Nano Lett. **15**, 6030–6035 (2015)
4. P. Luo, Z. Ji, C. Li, G. Shi, Nanoscale **5**, 7361–7367 (2013)
5. H. Ding, S.-B. Yu, J.-S. Wei, H.-M. Xiong, ACS Nano (2015)
6. F. Zhang, F. Liu, C. Wang, X. Xin, J. Liu, S. Guo, J. Zhang, ACS Appl. Mater. Interfaces **8**, 2104–2110 (2016)
7. J.C. Vinci, I.M. Ferrer, S.J. Seedhouse, A.K. Bourdon, J.M. Reynard, B.A. Foster, F.V. Bright, L.A. Colón, J. Phys. Chem. Lett. **4**, 239–243 (2013)
8. L. Zheng, Y. Chi, Y. Dong, J. Lin, B. Wang, J. Am. Chem. Soc. **131**, 4564–4565 (2009)
9. L.-L. Li, J. Ji, R. Fei, C.-Z. Wang, Q. Lu, J.-R. Zhang, L.-P. Jiang, J.-J. Zhu, A facile microwave avenue to electrochemiluminescent two-color graphene quantum dots. Adv. Funct. Mater. **22**, 2971–2979 (2012)
10. P. Zhang, Z. Xue, D. Luo, W. Yu, Z. Guo, T. Wang, Anal. Chem. **86**, 5620–5623 (2014)
11. L. Zhao, F. Di, D. Wang, L.-H. Guo, Y. Yang, B. Wan, H. Zhang, Nanoscale **5**, 2655–2658 (2013)
12. S. Nandi, R. Malishev, K. Parambath Kootery, Y. Mirsky, S. Kolusheva, R. Jelinek, Chem. Commun. **50**, 10299–10302 (2014)
13. Y.P. Sun et al., J. Am. Chem. Soc. **128**, 7756–7757 (2006)
14. X. Wang et al., Chem. Commun. 3774–3776 (2009)
15. Ding et al., ACS Nano **10**, 484–491 (2016)

Chapter 4
Biological Applications of Carbon-Dots

As Carbon-Dots are biocompatible and readily functionalized nanoparticles, they have attracted significant interest as conduits for biologically oriented applications. The remarkable fluorescence properties of Carbon-Dots have naturally been primary targets for biological research, particularly focusing on *bio-imaging* applications (discussed in more detail in Chap. 5). Carbon-Dots have been employed, however, in biological applications other than imaging, mostly related to their nanoscale dimensionalities and the functional units displayed upon their surface. Several examples of such applications are discussed below. It should be emphasized that biological applications of Carbon-Dots still contain many "mechanistic unknowns," somewhat related to the fact that Carbon-Dot research is still a young scientific discipline.

Several reports have described applications of Carbon-Dots as potential *therapeutic agents* to different diseases. S. Shankar and colleagues at CSIR-NIIST, India, have discovered that green-luminescent Carbon-Dots prepared from styrene soot had an intriguing anti-angiogenic effect [1]. Angiogenesis is the physiological process associated with many tumors in which cancer cells induce proliferation of blood vessels around the tumor, thereby providing the needed nutrients for rapid multiplication of the cells and tumor growth. Indeed, blocking angiogenesis is considered a promising route in anticancer drug development. Interestingly, the researchers have shown that addition of Carbon-Dots significantly reduced angiogenesis in an in vitro assay (Fig. 4.1). While the precise mechanism for angiogenesis inhibition by the Carbon-Dots has not been determined, the researchers discovered that addition of the Carbon-Dots resulted in lower expression of important cellular growth factors known to promote blood vessel growth, leading to the hypothesis that the carbon nanoparticles blocked growth factor activity through binding to these proteins.

The biocompatibility of Carbon-Dots combined with the broad variety of surface functionalization pathways have opened avenues for *gene delivery* applications. Many currently-employed gene carriers utilize positively charged polymeric materials as conduits for delivery of (negatively charged) DNA fragments. W. Liu and

© Springer International Publishing Switzerland 2017

R. Jelinek, *Carbon Quantum Dots*, Carbon Nanostructures,
DOI 10.1007/978-3-319-43911-2_4

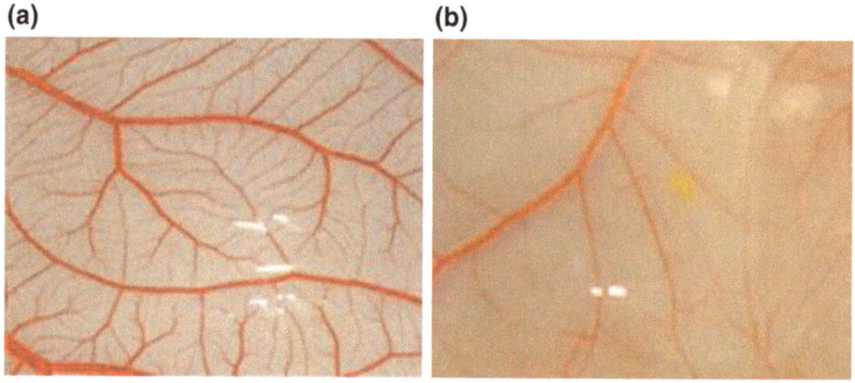

Fig. 4.1 Anti-angiogenic effect of carbon-dots. **a** Vascular density observed in buffer (recorded for chick embryos in a chick chorioallantoic membrane (CAM) assay); **b** much lower vascular proliferation upon treatment with Carbon-Dots. Reprinted with permission from Shereema et al. [1]. Copyright (2015) American Chemical Society

colleagues at Tianjin University, China, have utilized such a positively charged polymer—polyethylenimine (PEI)—as the starting material (i.e., carbon source) for preparation of a Carbon-Dot gene delivery platform [2]. Hydrothermal treatment of the carbonaceous building blocks resulted in the formation of graphitic core of the Carbon-Dots; however, as observed in numerous Carbon-Dot systems—the *surfaces* of the dots display the functional units within the starting reagents. This scenario was also apparent in the PEI-generated Carbon-Dot system (Fig. 4.2). The positive moieties upon the Carbon-Dot surface effectively immobilized the DNA fragments, delivering the Carbon-Dot/DNA complexes into cells. Particularly important in the context of gene delivery applications, the transfected cells appeared healthy and the transfection efficiency was on par with conventional gene carriers.

While the system described above accomplished DNA delivery through associating the DNA cargo with the Carbon-Dots through electrostatic attraction, other studies have demonstrated *direct synthesis* of Carbon-Dots from oligonucleotides as the carbon source, and subsequent cell uptake of the dots. Figure 4.3 illustrates such an approach, in which purified natural DNA was hydrothermally treated, yielding fluorescent Carbon-Dots. In this work, Z. Chen and colleagues at Jilin University, China, prepared DNA-extracted Carbon-Dots and showed that they could be readily internalized into cells [3]. While these Carbon-Dots have not been employed yet as a platform for transporting *genes* into cells, the researchers did demonstrate successful delivery of varied drug compounds into bacterial and mammalian cells, underscoring the versatility of the DNA-prepared Carbon-Dots as a drug delivery vehicle.

Carbon-Dots have been employed as scaffolds for intracellular delivery of other molecular cargoes, such as small-interfering RNA (siRNA, also known as "silencing RNA"). siRNA technology is a promising strategy for turning off expression of target genes; however, transporting the (negatively charged) siRNA across the cell membrane is among the most formidable challenges for practical applications of the

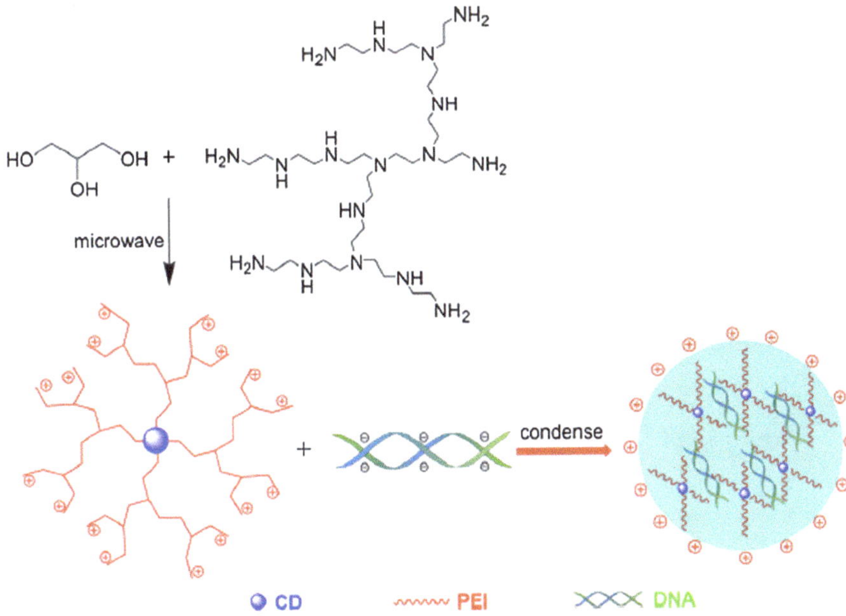

Fig. 4.2 Carbon-dots as vehicles for gene delivery. Scheme depicting preparation of the gene delivery platform. The Carbon-Dots are prepared through microwave-induced hydrothermal treatment of glycerol and polyethylenimine (PEI). The as-synthesized Carbon-Dots were incubated with DNA, forming the transfection agent upon further condensation. Reprinted from Liu et al. [2], Copyright (2012), with permission from Elsevier

technology. Figure 4.4 illustrates a siRNA cell delivery vehicle based upon Carbon-Dots [4]. The siRNA transport system, designed by D. Cui and colleagues in Shanghai Jiao Tong University, China, relied upon coating the Carbon-Dots with polyethylenimine (PEI), accomplished through electrostatic attraction between the positively charged PEI and the Carbon-Dot surface (which displays abundant carboxylate moieties). PEI had a dual role in the experiment. First, PEI—a widely used cell transfection agent—enabled transport across the membrane and uptake of the Carbon-Dot complexes into the cell. In addition, PEI functioned as a scaffold for docking of the siRNA molecules through electrostatic affinity to the positively charged particles' surface. Importantly, the siRNA-PEI-Carbon-Dot complex retained an overall positive charge which is essential for facilitating cell uptake.

The functional assay data depicted in Fig. 4.4b confirmed the physiological effectiveness of the siRNA/Carbon-Dot platform. In this experiment, the researchers incorporated within the Carbon-Dots siRNA targeting the *Survivin* gene which inhibits apoptotic (e.g., cell death) processes. Indeed, the flow cytometry analysis in Fig. 4.4b reveals that treatment with the Carbon-Dots coupled to siRNA designed to downregulate Survivin gave rise to *more pronounced* apoptotic processes, manifested as more abundant apoptotic cells *outside* of the bottom-left quadrant (which corresponds to healthy cells).

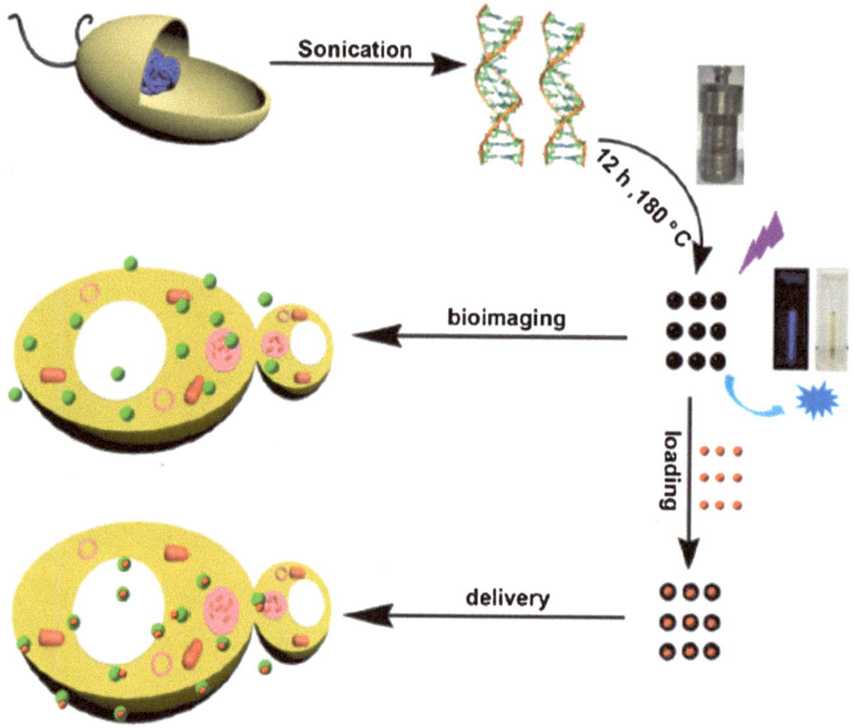

Fig. 4.3 Carbon-dots prepared from oligonucleotides as a cellular delivery vehicle. The scheme depicts the hydrothermal synthesis of Carbon-Dots from purified DNA, and employing the Carbon-Dots for either bioimaging (making use of their fluorescence), or delivery into cells (upon attachment of molecular cargo onto the Carbon-Dots). Reprinted with permission from Ding et al. [3]. Copyright (2015) American Chemical Society

An important advantage of Carbon-Dots as a cellular delivery vehicle is the possibility to combine transport of the molecular cargo and *imaging*—employing the *luminescence* properties of the Carbon-Dots—thus enabling assessment of the effectiveness of delivery. Many examples of combined Carbon-Dot-assisted drug delivery/target visualization have been reported. Figure 4.5 presents an elegant multifunctional system in which Carbon-Dots served both as a platform for targeted delivery of a therapeutic agent to intracellular organelles (mitochondria), as well as an imaging constituent. Specifically, the Carbon-Dots prepared by S. Wu and colleagues at South China University of Technology, displayed two important molecular elements: a *mitochondria-recognition* moiety (triphenylphosphonium, TPP), and a nitric oxide (NO)—releasing moiety (NO blocks mitochondria respiration and is thus a potent cytotoxic agent) [5]. The researchers showed that the functionalized Carbon-Dots had anticancer activity through binding intracellular mitochondria (via the mitochondria targeting moiety upon the Carbon-Dots' surface), inducing mitochondria damage (through release of NO upon light irradiation

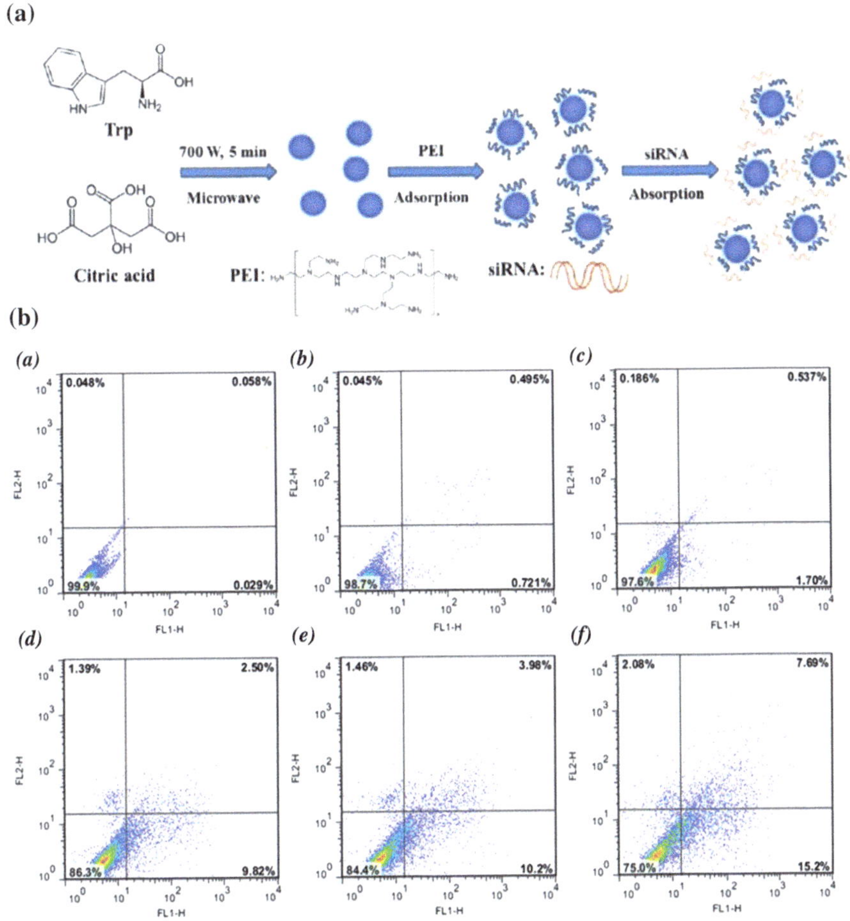

Fig. 4.4 Carbon-dots as a delivery vehicle for small-interfering RNA (siRNA). **a** Scheme illustrating the preparation procedure. The Carbon-Dots were coated with a positively charged polymer (PEI) which was employed for docking siRNA through electrostatic attraction. **b** Flow cytometry data illuminating the effect of siRNA/Carbon-Dots targeting the Survivin gene. *a–c* control or mock-infected cells. Almost all cells are located in the *lower left* quadrant, corresponding to non-apoptotic cells. *d–f* cells treated with different siRNA/Carbon-Dot formulations. In all cases significantly greater fractions of cells are early apoptotic (*lower right* quadrant) and late-apoptotic (*upper right* quadrant). Reprinted with permission from Wang et al. [4]. BioMed Central (2014)

of the cells), and consequent apoptosis of the cancer cells (Fig. 4.5). Importantly, the fluorescence properties of the Carbon-Dots were also exploited to assess the localization and accumulation of the particles in the mitochondria.

An interesting application in which *hollow* Carbon-Dots were synthesized, designed to act as a host assembly for drug cargoes, is shown in Fig. 4.6. The hollow dots were prepared by H. Zheng and colleagues at Southwest University,

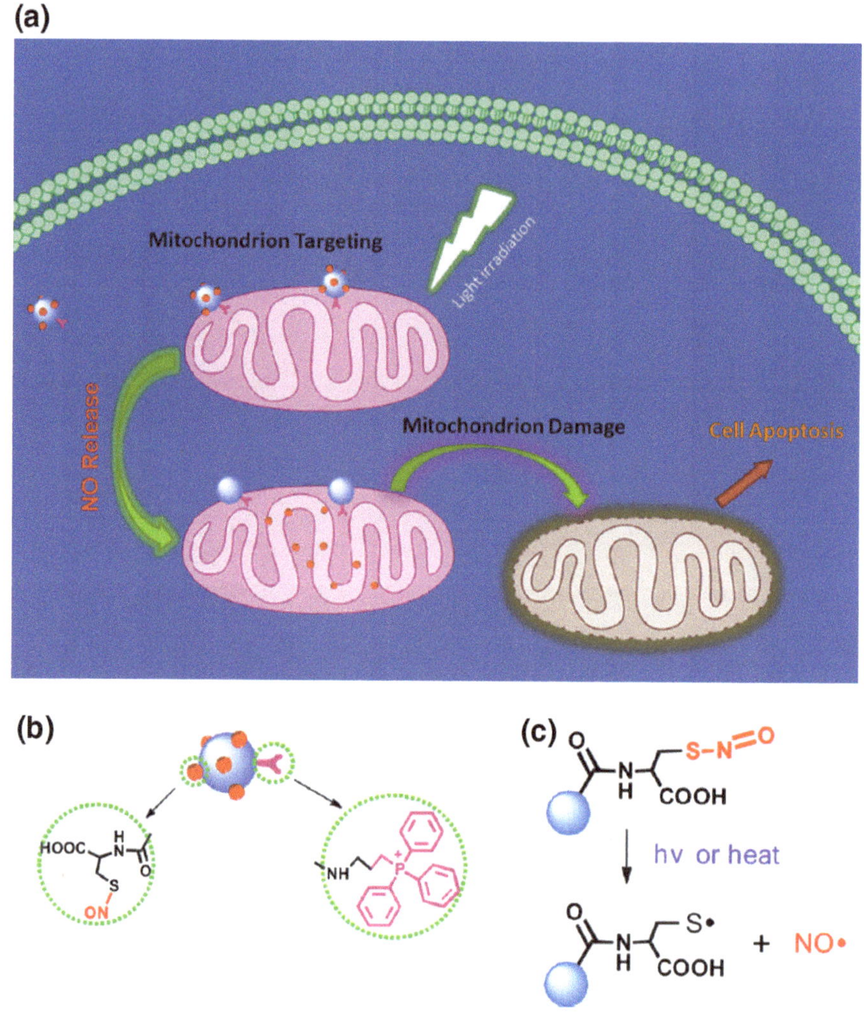

Fig. 4.5 Multifunctional mitochondria targeting and cytotoxic carbon-dots. **a** Schematic drawing of the Carbon-Dots action. The Carbon-Dots are targeted to mitochondria through surface-display of a mitochondria-recognition element (triphenylphosphonium, TPP); NO is then released upon light irradiation, resulting in mitochondria damage and cell apoptosis. **b** Structure of the Carbon-Dots, highlighting the two functional elements—S-nitrosothiol, the NO release moiety and TPP. **c** Mechanism of light-induced release of the NO radical from S-nitrosothiol. Reprinted from Xu et al. [5], with permission from John Wiley & Sons (2014)

China, through a two-step carbonization process utilizing bovine serum albumin, a common blood protein, as the carbon source [6]. The researchers demonstrated that the hollow Carbon-Dots could upload a desired anticancer drug compound, and successfully deliver it into cells. Importantly, the hollow Carbon-Dots retained their luminescence properties in the "loaded" state, thus constituting a multifunctional

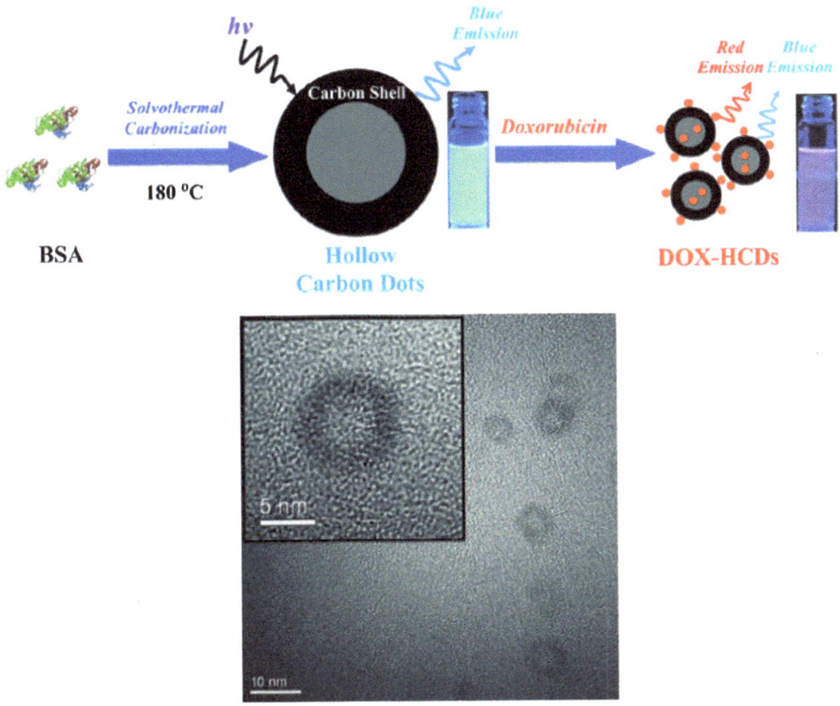

Fig. 4.6 Hollow carbon-dots for drug delivery. *Top* illustration of the preparation procedure of the hollow Carbon-Dots from bovine serum albumin (BSA), and uploading of the drug cargo (the anticancer substance doxorubicin). The DOX/Carbon-Dot composite particles emitted both in *blue* (The Carbon-Dots' fluorescence) and *red* (DOX fluorescence). *Bottom* high resolution transmission electron microscopy (HRTEM) image of the hollow Carbon-Dots. Reprinted from Liu et al. [2], Copyright (2012), with permission from Elsevier

platform, both as a drug delivery vehicle, and enabling tracking of the particles using fluorescence microscopy.

An application of Carbon-Dots for drug delivery which also made use of their luminescence properties is presented in Fig. 4.7. In this study, K. Sreenivasaan and colleagues at Sree Chitra Tirunal Institute for Medical Sciences and Technology, India, demonstrated the use of functionalized Carbon-Dots for both imaging of bone cracks as well as delivery of therapeutic substances to the crack area [7]. To achieve these goals, the researchers covalently linked the Carbon-Dots to *glutamic acid*, an amino acid exhibiting affinity to calcium, and *ciproflaxin*, an antibacterial drug. As shown in Fig. 4.7, localization of the fluorescence signal around the crack area within the bone was observed, indicating effective targeting of the Carbon-Dots. The researchers also reported that the Carbon-Dots exhibited antibacterial effects, confirming delivery and release of ciproflaxin. A significant question arising from this study and therapeutic applications of Carbon-Dots in general, however, is whether this strategy will be effective and useful also in vivo.

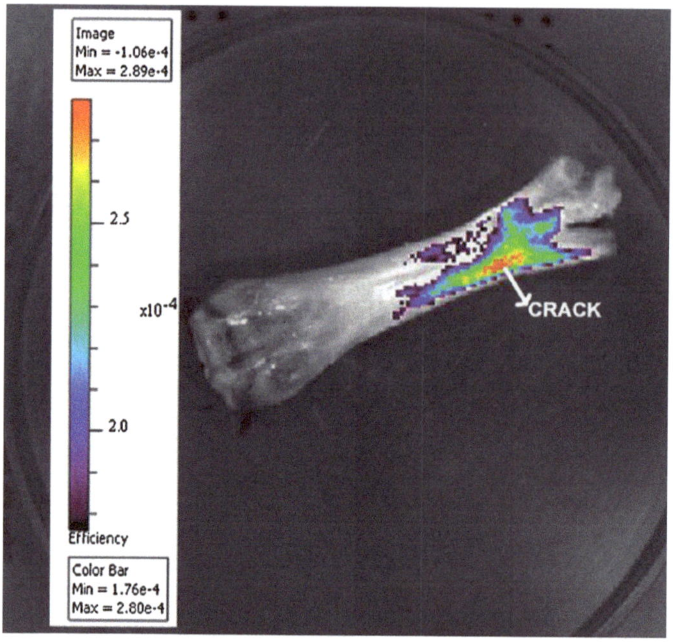

Fig. 4.7 Bone crack detection using calcium-targeting carbon-dots. The Carbon-Dots were conjugated with glutamic acid which exhibits affinity to calcium. Greater fluorescence intensity at the crack area confirms localization of the Carbon-Dots. Reproduced from Krishna et al. [7], with permission of the Royal Society of Chemistry

Carbon-Dots have been touted as possible conduits for other therapeutic applications. A recent study has demonstrated application of Carbon-Dots as *scavengers* of reactive oxygen species (ROS). ROS display unpaired electrons and are considered major toxic substances and risk factors in varied diseases. S. Dhara and colleagues at the Indian Institute of Technology, Kharagpur, discovered that Carbon-Dots synthesized from molasses exhibited anti-oxidant properties—reducing concentrations of active radicals in vitro. This phenomenon was explained by reactions of proton-donors at the Carbon-Dot surface with the electron-contributing ROS, ultimately resulting in migration of the electrons to the Carbon-Dots' carbon cores (Fig. 4.8).

Carbon-Dot conjugates have been proposed as platforms for *photodynamic therapy* (*PDT*). PDT is based upon targeting a singlet oxygen-releasing "photosensitizer" (PS) onto tumors; the oxygen radical is subsequently released through light irradiation of the PS, destroying tumor cells in its environment. The photo-physical challenges in PDT mainly concern the efficiency of light (energy) absorbance by the PS; this can be a significant problem since human tissues absorb light in the blue and visible regions thereby restricting the available excitation window of the PS. Figure 4.9 illustrates a Carbon-Dot-based strategy for PDT

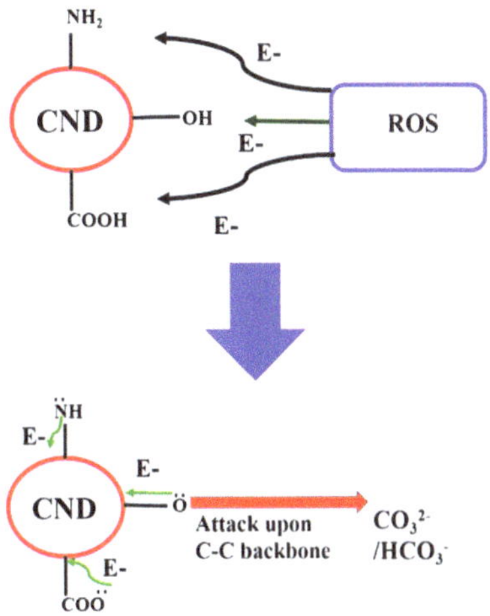

Fig. 4.8 Scavenging of reactive oxygen species (ROS) by carbon-dots. Proposed mechanism for the reaction of ROS with Carbon-Dots (termed "carbon nanodots," CND). Electrons are transferred from the ROS molecule to proton-donating residues upon the Carbon-Dot's surface, thereby reducing the toxicity of the ROS radicals. Reproduced from Das et al. [12], with permission of the Royal Society of Chemistry

applications, based upon *two-photon absorption* by the dots. The system, developed by J.Y. Chen and colleagues at Fudan University, China, comprised of Carbon-Dots covalently coupled to a porphyrin derivative (TMPyP, Fig. 4.9a) which is a known PS. TMPyP, however, mostly absorbs light in the *visible* range, thus severely restricting its potential therapeutic applicability. As depicted in Fig. 4.9, the designed Carbon-Dots could absorb light in near IR (700 nm), which, crucially,

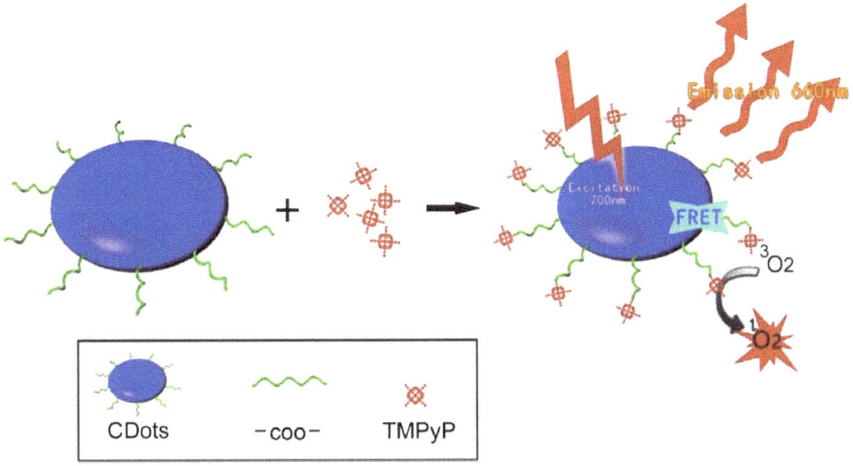

Fig. 4.9 Carbon-dots for photodynamic therapy (PDT). The Carbon-Dots were coupled to TMPyM (the photosensitizer). The spectral properties of the Carbon-Dots and TMPyP reveal overlap between the Carbon-Dots' emission and TMPyP excitation, enabling energy transfer from the Carbon-Dots, irradiated with near infrared (NIR) light, to the TMPyP. The photodynamic process is illustrated on the right, in which triplet oxygen employed for cell killing is generated. Reprinted from Wang et al. [13], Copyright (2014), with permission from Elsevier

penetrates through tissues. Through a two-photon process, a *shorter wavelength* light was emitted by the Carbon-Dots which coincided with the excitation wavelength of TMPyP making possible energy transfer through a Forster resonance energy transfer (FRET) process. The combination of the NIR light absorbance by the Carbon-Dots and efficient FRET enabled cell killing through triplet oxygen release by the TMPyP.

An innovative use of graphene quantum dots (GQDs) as PDT vehicles has been recently reported by J. Ge and colleagues at the Technical Institute of Physics and Chemistry, Beijing. The researchers showed that GQDs could generate the cytotoxic species *singlet oxygen*, presumably through transfer of the excitation energy of the dots to an "excited triplet state" [8]. According to the proposed model (Fig. 4.10), energy required for generating singlet oxygens could be produced via both transitions from the excited singlet state to the triplet state of the GQDs, as well as through transition from the triplet state to the ground state of the dots. The proposed "multistate sensitization" mechanism stands in contrast to the "single transition" photosensitization in conventional PDT agents and might account for the excellent sensitization efficiency, generating high concentrations of singlet oxygen by the GQDs. Importantly, GQDs offer several advantages as a PDT platform, since these nanoparticles are water-soluble and can be readily internalized in tumors. Furthermore, the *luminescence* properties of the GQDs can be simultaneously employed for cellular and tumor imaging, aiding to target the dots onto cells and tissues.

Carbon-Dots (specifically *graphene* quantum dots, GQDs) have been also shown to exhibit *antibacterial* properties. L. Yang and colleagues at the University of Science and Technology of China prepared GQDs from different sources—through

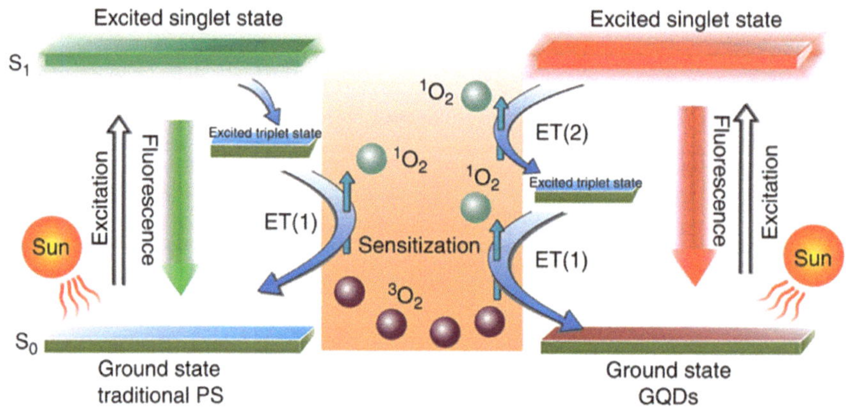

Fig. 4.10 Multisensitization model for photodynamic therapy of graphene quantum dots (GQDs). *Two* energy pathways for energy generation in GQDs (*right*) yield singlet oxygen more effectively, as compared to traditional photosensitizers which rely on a single energy transition (*left*). Reprinted by permission from Macmillan Publishers Ltd: Ge et al. [8] Copyright (2014)

chipping of C_{60} (buckminsterfullerene), and by cleavage of graphene oxide sheaths—and investigated their effects upon viabilities of different bacterial strains [9]. Interestingly, the experiments indicated that the GQDs had selective antibacterial effects, killing some bacterial species while not affecting others. The authors proposed an intriguing model in which the bactericidal effect is ascribed to disruption of the cell surface by the GQDs; accordingly, the *curvature match/mismatch* between the carbon nanoparticles and bacterial cells is the crucial determinant for GQD-cell wall interactions. Based upon this model, the specific killing of *s. aureus* by the GQDs prepared from C_{60} was ascribed to the similarity between the carbon-dots' curvature (as they were generated from the spherical C_{60} molecules) and the spherical *S. aureus* cells. Likewise, the GQDs produced from *graphene oxide* retained the planar, sheet-like structure thus the curvature mismatch minimized their antibacterial properties against *S. aureus* (Fig. 4.11). Similar relationship between curvature match/mismatch and cell killing was also recorded in case of the rod-shape *P. aeruginosa* bacteria (Fig. 4.11). In a broader context, this work might cast doubt on the premise that Carbon-Dots (and GQDs) are inherently *non-cytotoxic* as the carbon nanoparticles might still exhibit membrane disruption activity even towards mammalian cells.

Interactions of Carbon-Dots with *oligonucleotides* (i.e., DNA, RNA) could usher a plethora of new applications. In general, attachment of DNA or RNA to Carbon-Dots could be accomplished when Carbon-Dots display positive surface charge (as oligonucleotides are negative in physiological conditions). Figure 4.12

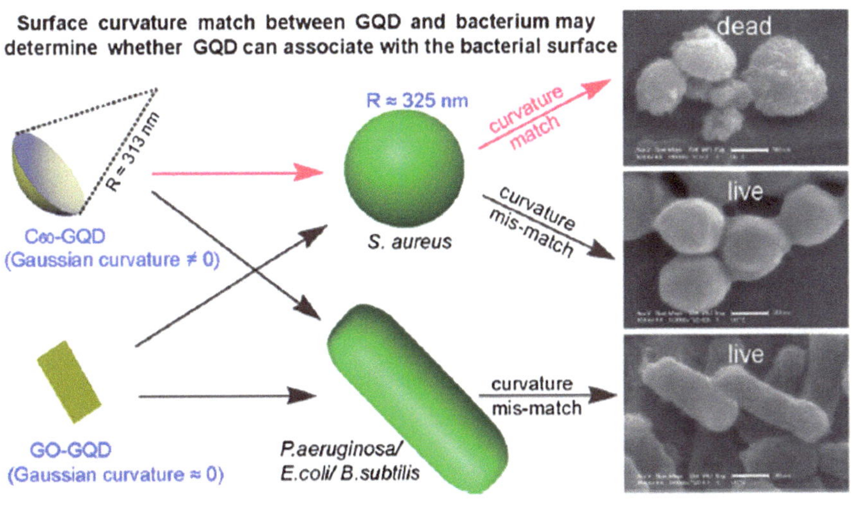

Fig. 4.11 Shape-dependent antibacterial properties of graphene quantum dots. GQDs produced from C_{60} exhibit antibacterial activity against *S. aureus* due to *curvature match* between the dots and the bacterial cells. Curvature *mis*match between *S. aureus* cells and GQDs synthesized from *graphene oxide* might explain their lack of antibacterial action. Similar considerations could explain the minimal antibacterial effects of both GQDs against *P. aeruginosa*. Reprinted with permission from Hui et al. [9]. Copyright (2016) American Chemical Society

illustrates a "logic gate" application of positively charged Carbon-Dots produced from glucose and spermine as the carbon sources. The research project, reported by X. Qu and colleagues at Changchun Institute of Applied Chemistry, China, demonstrated that the spermine-functionalized Carbon-Dots were specifically attracted to guanine-cytosine (GC) nucleotide sequences, and furthermore induced conformational transformation of DNA strands from the canonical *right handed* helix into *left handed helix* (termed "Z-DNA") [10]. The authors ascribed this interesting observation to the occurrence of electrostatic binding between the Carbon-Dots (specifically the positive spermine moieties) and the major groove of the DNA.

The logic operations demonstrated in the Carbon-Dot/DNA system are depicted in Fig. 4.12. Figure 4.12a presents the binding mode of the Carbon-Dots to the major groove and the different inputs available. Specifically, the two inputs were the *DNA strand* (a physiologically obtained oligonucleotide containing high abundance of GC pairs) and *ethidium bromide* (*EB*), a widely used fluorescent DNA binder which intercalates into the minor groove of double-stranded DNA. Importantly, the experiment was carried out using excitation of 400 nm, which excites *only* the Carbon-Dots, emitting at 465 nm. The different *outputs* are outlined

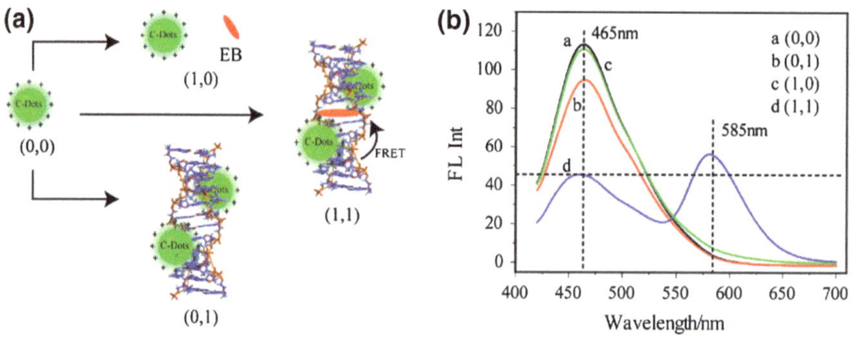

(c)

Input 1 [ct-DNA]	Input 2 [EB]	Output 1 λem=585nm	Output 2 λem=465nm
0	0	0	1
0	1	0	1
1	0	0	1
1	1	1	0

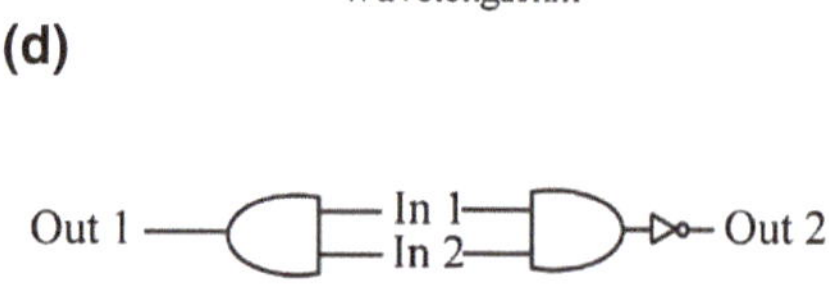

Fig. 4.12 Logic operations using DNA/carbon-dot assembly. **a** Illustration of the different inputs. (0,0)—neither DNA nor ethidium bromide (EB) are present; (1,0)—only EB present; (0,1)—only DNA present; (1,1)—EB and DNA are both present. The Carbon-Dots are added to the solutions in all cases. **b** Spectral features associated with the four different input combinations. The peak at 585 nm appears only in the (1,1) input, corresponding to FRET from Carbon-Dots to EB; in all other cases the Carbon-Dot emission at 465 nm is observed. **c** The logic table based upon the fluorescence emission spectra. **d** The AND (*left*) and NAND (*right*) logic operations. Reprinted from Feng et al. [10], by permission of Oxford University Press

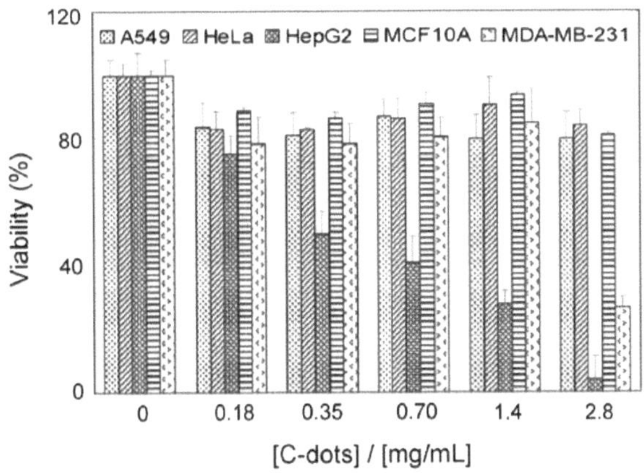

Fig. 4.13 Cancer cell death induced by carbon-dots produced from ginger. Viability of different cell lines after 24-h incubation with the Carbon-Dots. Dramatic toxic effect is apparent in case of HepG2 cells. Reproduced from Li et al. [11], with permission of the Royal Society of Chemistry

in Fig. 4.12b. In particular, the peak at 585 nm appeared *only* when both inputs were provided (i.e., DNA *and* EB) due to Forster resonance energy transfer (FRET) from the Carbon-Dots to EB. Accordingly, the fluorescence emission signal at 585 nm corresponded to the *AND* logic operation (generated when the *two* input events are provided). Similarly, the fluorescence peak at 465 nm (corresponding to the emission of Carbon-Dots upon excitation at 400 nm) was indicative of a *NAND* logic action (i.e., *no output*, reflecting a situation in which the two inputs are *not* present simultaneously).

Biocompatibility has naturally been among the attractive properties of Carbon-Dots in the context of biological applications. Some studies, however, have reported the use of Carbon-Dots as selective *toxic factors* for cancer cells. H.T. Chang and colleagues at the National Taiwan University demonstrated that Carbon-Dots prepared from *ginger juice* as the carbon source exhibited dramatic toxic effect upon human hepatocellular carcinoma (HepG2) cells [11]. As shown in Fig. 4.13, the toxicity of the Carbon-Dots was highly specific, as other types of cells (including *other* cancer cell types) appeared viable and un-perturbed upon addition of the Carbon-Dots. This intriguing result was ascribed to increased uptake of the Carbon-Dots by HepG2 cells (possibly promoted by specific ginger-originated compounds upon the Carbon-Dots' surface). The cells subsequently underwent apoptosis associated with enhanced intracellular levels of reactive oxygen species (ROS), directly induced by the cell-internalized Carbon-Dots. The researchers have further demonstrated the therapeutic potential of the Carbon-Dots in significantly reducing the sizes of actual tumors in mice.

References

1. R.M. Shereema, T.V. Sruthi, V.B.S. Kumar, T.P. Rao, S.S. Shankar, Nano-carrier for gene delivery and bioimaging based on carbon-dots with PEI-passivation enhanced fluorescencetitle. Biochemistry **54**, 6352–6356 (2015)
2. C. Liu, P. Zhang, X. Zhai, F. Tian, W. Li, J. Yang, Y. Liu, H. Wang, W. Wang, W. Liu, Nano-carrier for gene delivery and bioimaging based on carbon dots with PEI-passivation enhanced fluorescencetitle. Biomaterials **33**, 3604–3613 (2012)
3. H. Ding, F. Du, P. Liu, Z. Chen, J. Shen, ACS Appl. Mater. Interfaces **7**, 6889–6897 (2015)
4. Q. Wang, C. Zhang, G. Shen, H. Liu, H. Fu, D. Cui, Fluorescent carbon dots as an efficient siRNA nanocarrier for its interference therapy in gastric cancer cells. J Nanobiotechnol. **12**, 58 (2014)
5. J. Xu, F. Zeng, H. Wu, C. Hu, C. Yu, S. Wu, Preparation of a mitochondria-targeted and NO-releasing nanoplatform and its enhanced pro-apoptotic effect on cancer cells. Small **10**, 3750–3760 (2014)
6. Q. Wang, X. Huang, Y. Long, X. Wang, H. Zhang, R. Zhu, L. Liang, P. Teng, H. Zheng, Carbon **59**, 192–199 (2013)
7. A.S. Krishna, C. Radhakumary, M. Antony, K. Sreenivasan, J. Mater. Chem. B **2**, 8626–8632 (2014)
8. J. Ge, M. Lan, B. Zhou, W. Liu, L. Guo, H. Wang, Q. Jia, G. Niu, X. Huang, H. Zhou, X. Meng, P. Wang, C.-S. Lee, W. Zhang, X. Han, Nat. Commun. **5** (2014)
9. L. Hui, J. Huang, G. Chen, Y. Zhu, L. Yang, ACS Appl. Mater. Interfaces **8**, 20–25 (2016)
10. L. Feng, A. Zhao, J. Ren, X. Qu, Lighting up left-handed Z-DNA: photoluminescent carbon dots induce DNA B to Z transition and perform DNA logic operations. Nucleic Acids Res. **41**, 7987–7996 (2013)
11. C.-L. Li, C.-M. Ou, C.-C. Huang, W.-C. Wu, Y.-P. Chen, T.-E. Lin, L.-C. Ho, C.-W. Wang, C.-C. Shih, H.-C. Zhou, Y.-C. Lee, W.-F. Tzeng, T.-J. Chiou, S.-T. Chu, J. Cang, H.-T. Chang, J. Mater. Chem. B **2**, 4564 (2014)
12. B. Das et al., J. Mater. Chem. B **2**, 6839 (2014)
13. J. Wang et al., Carbon nanodots featuring efficient FRET for two-photon photodynamic cancer therapy with a low fs laser power density. Biomaterials **35**, 9372–9381 (2014)

Chapter 5
Bioimaging Applications of Carbon-Dots

Similar to their *inorganic* nanoparticle relatives—semiconductor quantum dots (Q-dots)—Carbon-Dots have been recognized early on as conduits for biological imaging, particularly *cell* imaging. In this context, Carbon-Dots exhibit important fundamental and practical features that make them promising imaging materials. Specifically, Carbon-Dots can be engineered to possess bright luminescence (e.g., exhibit high quantum yield). Particularly important, the excitation-dependent emission properties encountered in many Carbon-Dot systems bestow *flexibility* for selection of excitation wavelengths in fluorescence microscopy experiments. Carbon-Dots, in fact, present some notable advantages as compared to Q-dots for bio-imaging (and other biological applications, as discussed above). Carbon-Dots have been mostly synthesized from natural carbon sources thus they are generally considered biocompatible and less toxic compared to inorganic Q-dots. Indeed, the diverse building blocks and numerous synthetic pathways make Carbon-Dots an inexpensive bio-imaging platform. In addition, the carbon scaffoldings of Carbon-Dots are amenable to diverse and facile functionalization routes providing ready means for surface-display of a variety of recognition elements for biological and cellular *targeting*. These aspects are highlighted in the representative examples of bioimaging applications of Carbon-Dots discussed below.

Figure 5.1 presents a typical multicolor imaging application of Carbon-Dots for cell imaging. In the experiment, cell labeling was accomplished using *amphiphilic Carbon-Dots*—Carbon-Dots comprising long hydrocarbon chains that were covalently attached to the graphitic cores [1]. The amphiphilic Carbon-Dots could be readily incorporated within lipid vesicles that were subsequently internalized by the cells, likely through endocytic pathways. The confocal fluorescence microscopy images in Fig. 5.1 confirm effective cell uptake of the Carbon-Dots and their uniform distribution within cellular environments including the cytosol and nucleoli (intra-nucleus organelles in eukaryotic cells). The fluorescence microscopy images in Fig. 5.1 also dramatically demonstrate the *multicolor* imaging capabilities accomplished using the Carbon-Dots, as images showing the distinctive features of the cells could be obtained through different excitation/emission wavelengths.

© Springer International Publishing Switzerland 2017
R. Jelinek, *Carbon Quantum Dots*, Carbon Nanostructures,
DOI 10.1007/978-3-319-43911-2_5

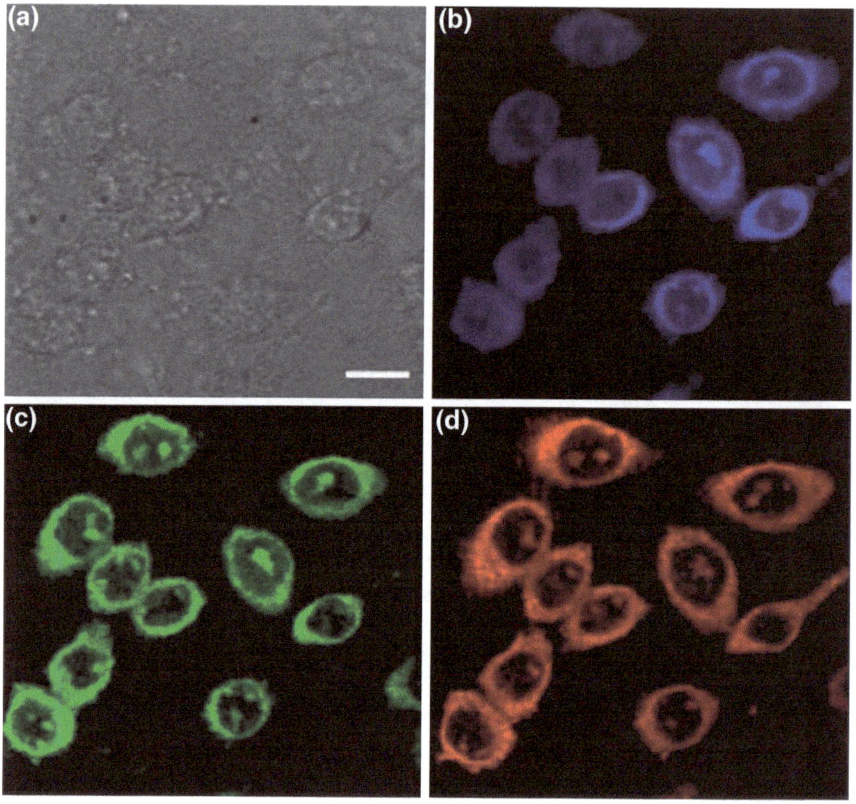

Fig. 5.1 Cell imaging with amphiphilic carbon dots. Bright-field image (**a**) and confocal fluorescence microscopy images of CHO cells incubated with amphiphilic Carbon-Dot embedded within phospholipid small unilamellar vesicles. The images were recorded at excitation of 405 nm emission filter 525/30 nm (**b**); excitation of 488 nm emission filter 525/30 nm (**c**); excitation at 561 nm emission 641/40 nm (**d**). *Scale bar* is 10 μm. Reproduced from Nandi et al. [1], with permission of the Royal Society of Chemistry

Furthermore, Carbon-Dot labeling appeared not to adversely affect cell viability as judged by the healthy shapes of the cells and their un-disrupted functionalities [1].

Cell uptake and staining can be accomplished also by using *hydrophilic* (rather than amphiphilic) Carbon-Dots. In fact, most Carbon-Dot-based bio-imaging applications reported thus far have been carried out by employing water-soluble Carbon-Dots. Figure 5.2 presents confocal fluorescence microscopy images of epithelial cells incubated with different Carbon-Dots prepared, respectively, from three phenylenediamine precursors which endowed each Carbon-Dot species with a distinct emission wavelength (i.e., distinct color) [2]. As shown in Fig. 5.2, the bright and monochromatic water-soluble Carbon-Dots, prepared by H. Lin and colleagues at the Ningbo Institute of Materials Technology and Engineering, China, were internalized by the cells, largely localized in the cytosol.

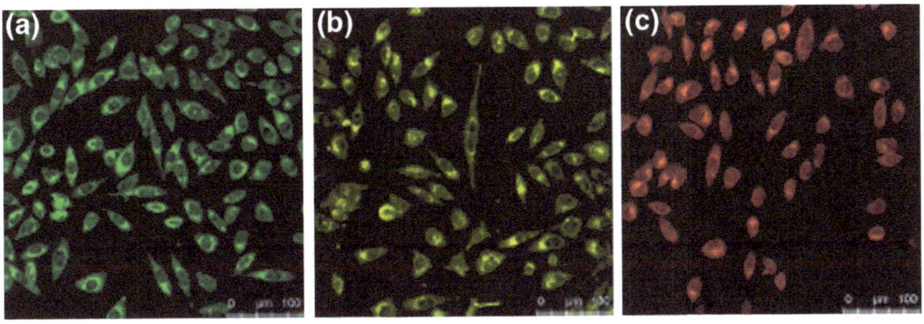

Fig. 5.2 Cell labeling with single-color carbon dots. Confocal fluorescence microscopy images (excitation at 405 nm) of MCF-7 cells incubated with Carbon-Dots prepared from *meta-* (**a**), *ortho-* (**b**), and *para-* (**c**) phenylenediamine. Each Carbon-Dot label provides a distinct fluorescence emission peak (e.g., distinct color). Reprinted from Jiang et al. [2], with permission from John Wiley & Sons (2015)

While most Carbon-Dot-based imaging experiments have used conventional routes of exciting the Carbon-Dots and recording their fluorescence emission, other innovative imaging strategies have been reported, employing fluorescence phenomena such as Forster resonance energy transfer (FRET). G. Zheng and colleagues at Fudan University, China, have demonstrated a FRET-based imaging approach in which Carbon-Dots functioned as energy *donors* and a drug compound (doxorubicin, *DOX*, a prominent anticancer therapeutic agent) was the energy *acceptor* [3]. Notably, the experiment enabled both microscopic visualization of the cell targets, as well as illuminating drug distribution and delivery. As shown in Fig. 5.3, the molecular system, serving both as a *drug delivery* vehicle and imaging platform, comprised of a Carbon-Dot coated with biologically inert polyethyleneglycol (PEG), further loaded with the DOX molecules. Microscopy imaging was facilitated through excitation of the Carbon-Dots using UV light, which resulted in FRET to the DOX molecules, consequently imaged through their green emission. Importantly, this FRET-based imaging scheme was applied for monitoring *release* of the drug molecules in cells. The time-dependent microscopy experiments in Fig. 5.3c nicely shows the gradual decay of the DOX fluorescence signals due to their release inside the cells.

Cell targeting through display of *recognition elements* on Carbon-Dots' surfaces have been widely used in imaging applications. While functionalization of Carbon-Dots' surfaces with recognition elements has been aided by the availability of varied chemical derivatization routes, in many cases the recognition elements actually comprised parts of the *carbon sources* (e.g., reagents) in Carbon-Dot synthesis, thereby obviating the need for post-synthesis coupling processes *after* the Carbon-Dots are synthesized. This approach might seem surprising, as the hydrothermal carbonization process essentially transforms the carbonaceous molecules into the graphitic cores of the Carbon-Dots; however, the residual constituents located at the carbon core's surface appear in many cases to retain their chemical and structural identities, enabling their use for docking onto cellular and tissue targets. Examples of both targeting strategies are presented below.

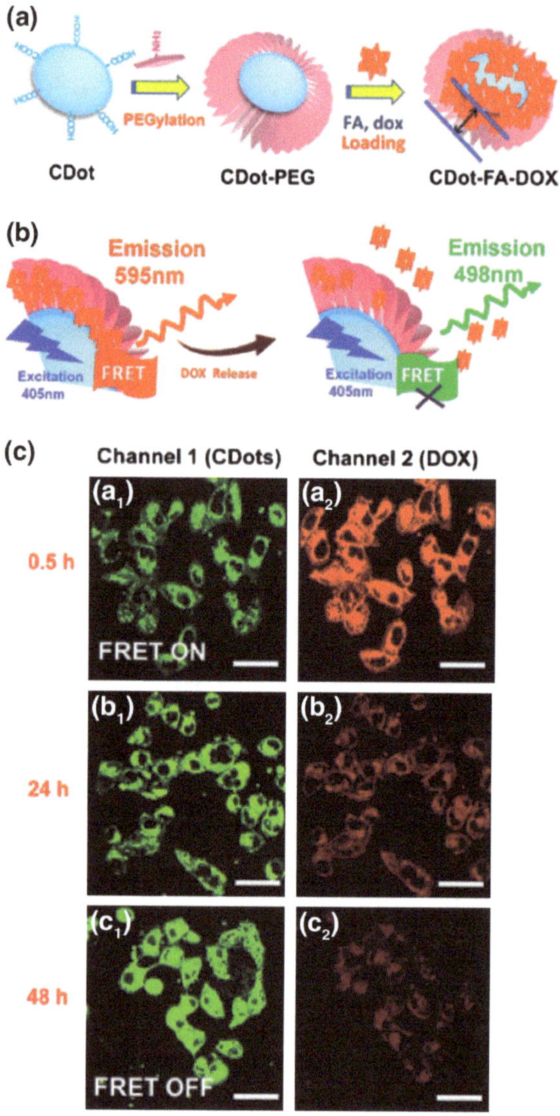

Fig. 5.3 Carbon dot assay for imaging drug delivery and distribution in cells. **a** Assembly scheme of the Carbon-Dots and doxorubicin (DOX). Polyethyleneglycol (PEG) and folic acid (FA) are employed as the coating layer to enhance biocompatibility and target cancer cells, respectively. Note the small distance maintained between the Carbon-Dot surface and DOX molecules, necessary to facilitate energy transfer from the Carbon-Dots. **b** Luminescence phenomena in the system: prior to DOX release, Forster resonance energy transfer (FRET) occurs between the Carbon-Dot and attached DOX molecules, generating luminescence at 595 nm (DOX emission). However, when the DOX cargo is released, the luminescence detected is from the Carbon-Dots (498 nm). **c** Confocal fluorescence microscopy images of HeLa cells treated with Carbon-Dots-DOX. Upon time progression a decrease of DOX fluorescence and simultaneous increase in Carbon-Dot fluorescence are observed. Reprinted from Tang et al. [3], with permission from John Wiley & Sons (2013)

Realization of cancer cell targeting through the use of *folic acid* as a recognition element is a case in point. The folate receptor is over-expressed upon surfaces of many types of cancer cells, making folic acid a useful targeting agent for those cells. Indeed, functionalization of therapeutic compounds and nanoparticles with folic acid has been an attractive approach for targeting such substances to cancer cells. Figure 5.4 depicts a strategy in which folic acid was linked to the surface of as-synthesized Carbon-Dots, which were subsequently employed for labeling of cancer cells [4]. Importantly, functionalization of Carbon-Dots with folic acid enabled targeting and fluorescence microscopy imaging of specific cancer cells (such as HeLa cells, Fig. 5.4b).

While the experiment outlined in Fig. 5.4 describes cancer cell targeting with Carbon-Dots that were *post-synthesis* functionalized with folic acid, similar labeling effect can be accomplished by production of Carbon-Dots from folic acid as the carbon source. Figure 5.5 demonstrates that Carbon-Dots, synthesized from *folic acid* in a simple, single-step hydrothermal treatment, served as an effective vehicle for imaging cancer cells expressing the folate receptor on their surface [5]. The Carbon-Dots exhibited clear cell selectivity which depended upon the relative

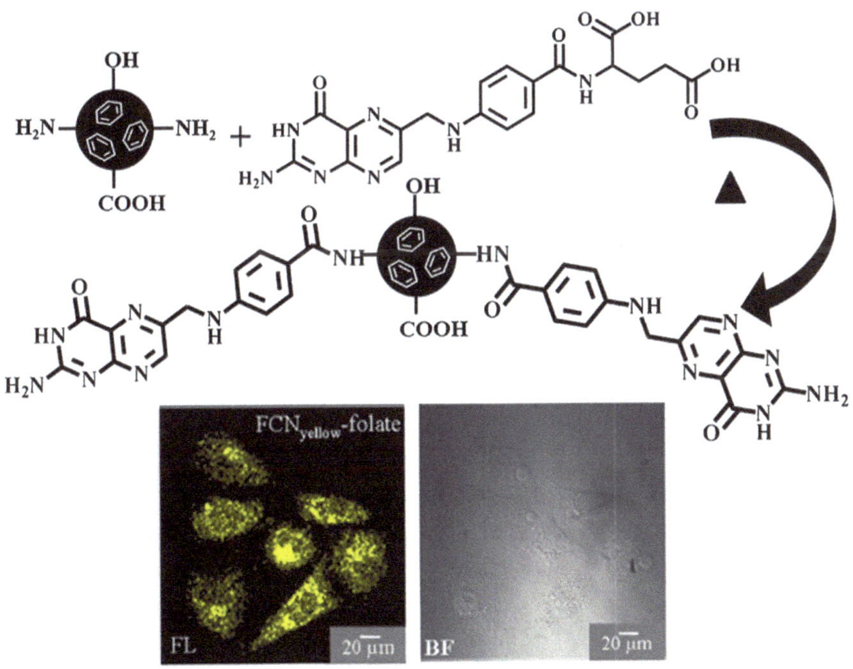

Fig. 5.4 Cancer cell imaging with carbon dots functionalized with folic acid. *Top* scheme of the preparation procedure, in which folic acid is covalently bonded to the as-synthesized Carbon-Dots. *Bottom* confocal fluorescence microscopy (*left*) and bright field (*right*) images of HeLa cells incubated with yellow Carbon-Dots (denoted "fluorescent carbon nanoparticles", FCN) functionalized with folic acid

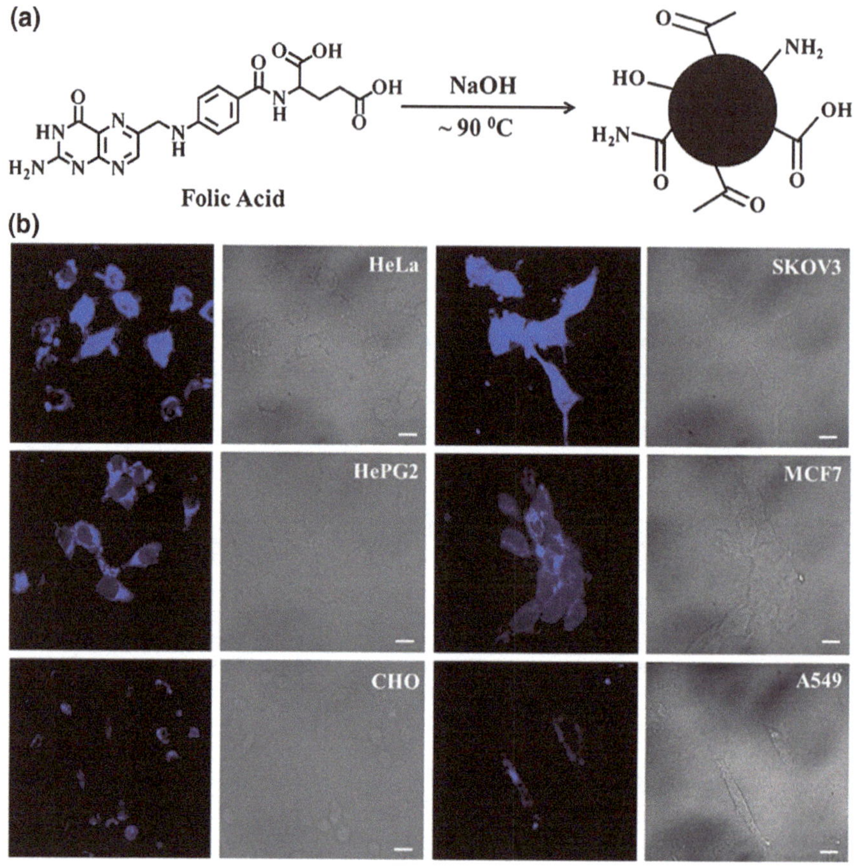

Fig. 5.5 Cancer cell imaging using carbon dots prepared from folic acid as the carbon source. **a** Synthesis scheme showing the single step formation process of the Carbon-Dots from folic acid; **b** confocal fluorescence images (exc. 405 nm, *left*) and bright field images (*right*) of different cells incubated with the folic acid-derived Carbon-Dots. Correlation is apparent between the degree of folate receptor expression upon the different cell surfaces and fluorescence labeling with the Carbon-Dots. Adapted from Bhunia et al. [5]

expression of the folate receptor (Fig. 5.5b), making such cells easily distinguishable in fluorescence microscopy imaging. Importantly, the extent of fluorescence labeling of the cancer cells could be modulated by co-addition of *free* folic acid to the cells (e.g., competition experiment), making possible identification of agonists/antagonists to the folate receptor and pointing to potential use of the Carbon-Dots as a drug development platform. The folic acid-derived Carbon-Dots were not cytotoxic thus might be employed in cancer bioimaging and diagnostics.

A notable property of Carbon-Dots directly pertinent to their use as bioimaging agents has been their applicability also in vivo. Specifically, Carbon-Dots exhibit fluorescence brightness, which could be retained for relatively long time periods in physiological environments such as tissues, and organs. Furthermore, in some instances Carbon-Dots emit sufficiently intense *longer wavelength* (*red*) *fluorescence* that is less absorbed by human tissues, thus allowing more effective imaging applications. These favorable photophysical properties are complemented by the biocompatibility and lower toxicity of Carbon-Dots, making them a potentially useful vehicle for in vivo imaging. Figure 5.6 presents an early study utilizing Carbon-Dots for in vivo imaging. In that work, Y.P. Sun and colleagues at Clemson University, US, prepared water-soluble Carbon-Dots which enabled in vivo imaging in mice following subcutaneous injection [6]. The bright luminescence made possible imaging of organs using different excitations/emissions (specifically–observing both in *green* fluorescence as well as *red* fluorescence). Notable from a practical standpoint, the Carbon-Dot emission did not decay even few hours after injection, underscoring the diagnostic potential of the dots.

An example of both in vitro and in vivo imaging of cancer cells and tumors is depicted in Fig. 5.7. In the experiments, Z. Sun and colleagues at Beijing University of Technology, China, synthesized Carbon-Dots from a mixture of D-glucose and L-aspartic acid [7]. These Carbon-Dots displayed enhanced selectivity and uptake toward cancerous *glioma* cells and tumors, presumably through specific binding between the aspartic acid/glucose moieties upon the Carbon-Dots and *integrins* (cell adhesion proteins) that are overexpressed in glioma cells. Notably, less pronounced Carbon-Dot targeting was observed when the researchers examined Carbon-Dots that were prepared from the *individual* constituents (D-glucose or L-aspartic acid) rather than their mixture, or when Carbon-Dots were synthesized from a mixture of glucose and *L-glutamic acid*—an amino acid with structural similarity to aspartic acid. These observations highlight the *specificity* of the carbon building blocks that can be retained even after Carbon-Dot synthesis. The in vivo experiment depicted in Fig. 5.7 (bottom) reveals that the Carbon-Dots specifically accumulated in the brain of glioma-bearing mice, indicating that the

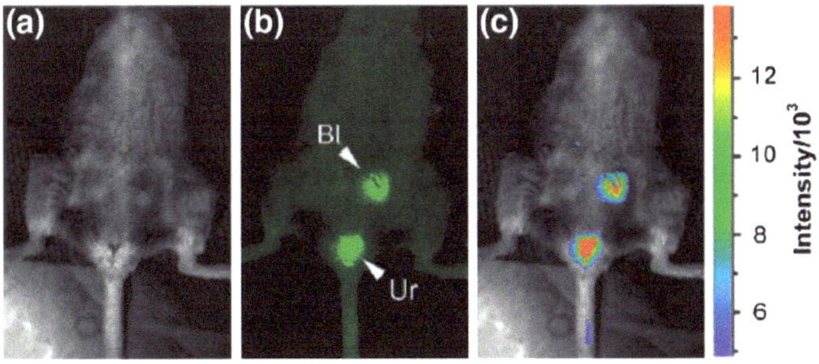

Fig. 5.6 In vivo fluorescence imaging with carbon dots. Carbon-Dots were intravenously injected to the mouse and imaged after 3 h. **a** bright field image; **b** fluorescence image (exc. 470 nm) showing accumulation of the Carbon-Dots in the bladder (Bl) and urine (Ur) tract; **c** color-coded fluorescence map illuminating distribution of the Carbon-Dots. Reprinted with permission from Yang et al. [6]. Copyright (2009) American Chemical Society

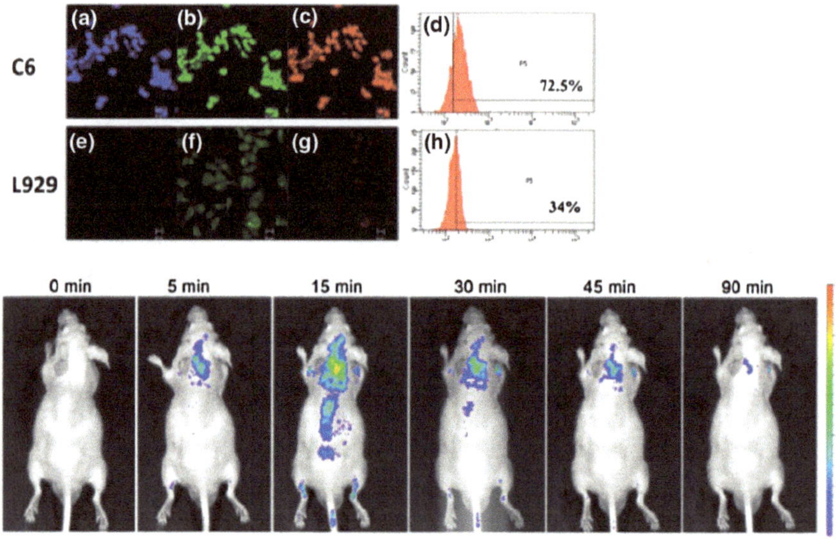

Fig. 5.7 In vitro and in vivo targeting of carbon dots prepared from specific ligands as carbon sources. The Carbon-Dots were synthesized from amino acid ligands targeting *glioma cancer cells*, presumably through binding to over-expressed integrins. *Top row* cell imaging experiments [exc. 405 nm (*a*, *e*), 488 nm (*b*, *f*), 555 nm (*c*, *g*)], and flow cytometry profiles (*d*, *h*) reflecting more significant labeling in case of cancerous C6 cells as compared to non-cancerous L929 cells. *Bottom row* whole-body in vivo fluorescence images of glioma-bearing mice intravenously–injected with Carbon-Dots confirming accumulation of Carbon-Dots in the brain and subsequent systemic clearance. Reprinted with permission from Zheng et al. [7]. Copyright (2015) American Chemical Society

carbon nanoparticles could be effectively transported across physiological barriers, particularly the blood–brain barrier (BBB). Overall, this study underscores the remarkable selectivity and imaging capabilities, both in vitro and in vivo, of Carbon-Dots that are simply prepared from molecular units which constitute ligands for specific cell receptors.

Several studies have reported on systems comprising Carbon-Dots conjugated with other imaging modalities. Figure 5.8, for example, schematically depicts the preparation of Carbon-Dots/*gadolinium ion* conjugates (Carbon-Dot-Gd nanoparticles). Gadolinium is a widely used contrast agent in magnetic resonance imaging (MRI), effective in reducing the T1 relaxation of water protons (thus enhancing their magnetic resonance signal as more scans can be conducted). The coupled Carbon-Dot-Gd particles were prepared by R. Zboril and colleagues at Palacky University, Czech Republic, through mixing and pyrolysis of the carbon source, a positively charged surface ligand, and a negatively charged Gd^{3+}-containing complex [8]. Following synthesis, the Gd^{3+} ions were likely embedded within the Carbon-Dot matrix, providing MRI capabilities in addition to the fluorescence imaging properties of the Carbon-Dots.

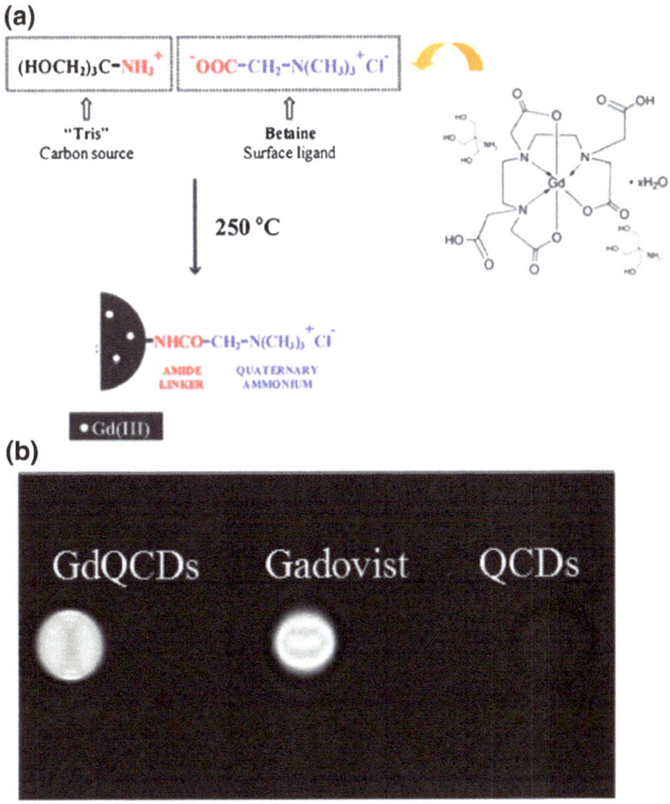

Fig. 5.8 Orthogonal imaging modalities of carbon dot composites. Carbon-Dots coupled with gadolinium ions for both fluorescence and magnetic resonance imaging. **a** Synthesis scheme depicting pyrolysis of a mixture comprising the carbon source, positive ligand, and Gd^{3+} complex. The gadolinium ions are embedded in the Carbon-Dot matrix. **b** T1-weighted magnetic resonance images (MRI) demonstrating significant brightness of the gadolinium-Carbon-Dot composites (GdQCDs), comparable to a commercially available MRI contrast agent (Gadovist). No MRI contrast apparent in case of Carbon-Dots not containing Gd^{3+}. Reproduced from Bourlinos et al. [8], with permission of the Royal Society of Chemistry

References

1. S. Nandi, R. Malishev, K. Parambath Kootery, Y. Mirsky, S. Kolusheva, R. Jelinek, Chem. Commun. **50**, 10299–10302 (2014)
2. K. Jiang, S. Sun, L. Zhang, Y. Lu, A. Wu, C. Cai, H. Lin, Red, green, and blue luminescence by carbon dots: full-color emission tuning and multicolor cellular imaging. Angew. Chem. Int. Ed. **54**, 5360–5363 (2015)
3. J. Tang, B. Kong, H. Wu, M. Xu, Y. Wang, Y. Wang, D. Zhao, G. Zheng, Carbon nanodots featuring efficient FRET for real-time monitoring of drug delivery and two-photon imaging. Adv. Mater. **25**, 6569–6574 (2013)
4. S.K. Bhunia, A. Saha, A.R. Maity, S.C. Ray, N.R. Jana, Sci. Rep. **3**, 1473 (2013)

5. S.K. Bhunia, A.R. Maity, S. Nandi, D. Stepensky, R. Jelinek, ChemBioChem **17**, 614–619 (2016)
6. S.-T. Yang, L. Cao, P.G. Luo, F. Lu, X. Wang, H. Wang, M.J. Meziani, Y. Liu, G. Qi, Y.-P. Sun, J. Am. Chem. Soc. **131**, 11308–11309 (2009)
7. M. Zheng, S. Ruan, S. Liu, T. Sun, D. Qu, H. Zhao, Z. Xie, H. Gao, X. Jing, Z. Sun, ACS Nano (2015)
8. A.B. Bourlinos, A. Bakandritsos, A. Kouloumpis, D. Gournis, M. Krysmann, E.P. Giannelis, K. Polakova, K. Safarova, K. Hola, R. Zboril, J. Mater. Chem. **22**, 23327–23330 (2012)

Chapter 6
Carbon-Dots in Sensing Applications

The photoluminescence (PL) properties of Carbon-Dots offer obvious avenues for sensing applications. In particular, well-known processes in fluorescent systems such as energy transfer, fluorescence quenching, and the sensitivity of fluorescent spectra to molecular environments, all point to broad applications of Carbon-Dots in bio- and chemo-sensing. Indeed, sensor applications of Carbon-Dots have been widely studied since the early days of this young field and diverse new sensing schemes have been reported in an ever-increasing rate. Representative Carbon-Dot-based sensors are discussed below, and for additional in-depth analysis the reader is referred to the many exhaustive reviews cited in the Introduction section. Thematically, this Chapter separately discusses Carbon-Dot *biosensors* and *chemosensors*; some overlap exists between the two topics, however the scientific and technological breadths of these two general applications make the separation logical from a didactic point of view.

6.1 Carbon-Dots in Biosensing

Some of the properties that make Carbon-Dots attractive in *biological* applications also play a role in their usefulness as a sensing platform. In particular, the perceived *biocompatibility* of Carbon-Dots is considered an added boon to biosensing applications, particularly in vivo, and accordingly Carbon-Dots have been proposed as promising conduits for physiological sensing. In addition, the facile strategies introduced for Carbon-Dot synthesis and functionalization facilitate display of varied functional units that can serve as recognition elements in biosensing applications. While many reports have presented the use of biomolecular recognition for *bioimaging*, this topic is discussed separately (Chap. 5), while here the focus is on biosensing applications in a more narrow perspective.

Most Carbon-Dot-based biosensing applications and mechanisms have focused on alteration of the *surface environments* of the dots, consequently modulating

© Springer International Publishing Switzerland 2017

R. Jelinek, *Carbon Quantum Dots*, Carbon Nanostructures,
DOI 10.1007/978-3-319-43911-2_6

the corresponding surface energy states and thus affecting the fluorescence emission—either the fluorescence *intensity* and/or the emission *shift*. Figure 6.1 provides an example of a Carbon-Dot system generating *intensity variations* of fluorescence emission induced by a target analyte. Specifically, the Carbon-Dots synthesized by Y. Tian and colleagues at Tongji University, China, exhibited dramatic response to the *solution pH* through change of the emission intensity of Carbon-Dots functionalized with proton receptors [1]. In the system depicted in Fig. 6.1, a "terpyridine"-based receptor molecule covalently linked onto the

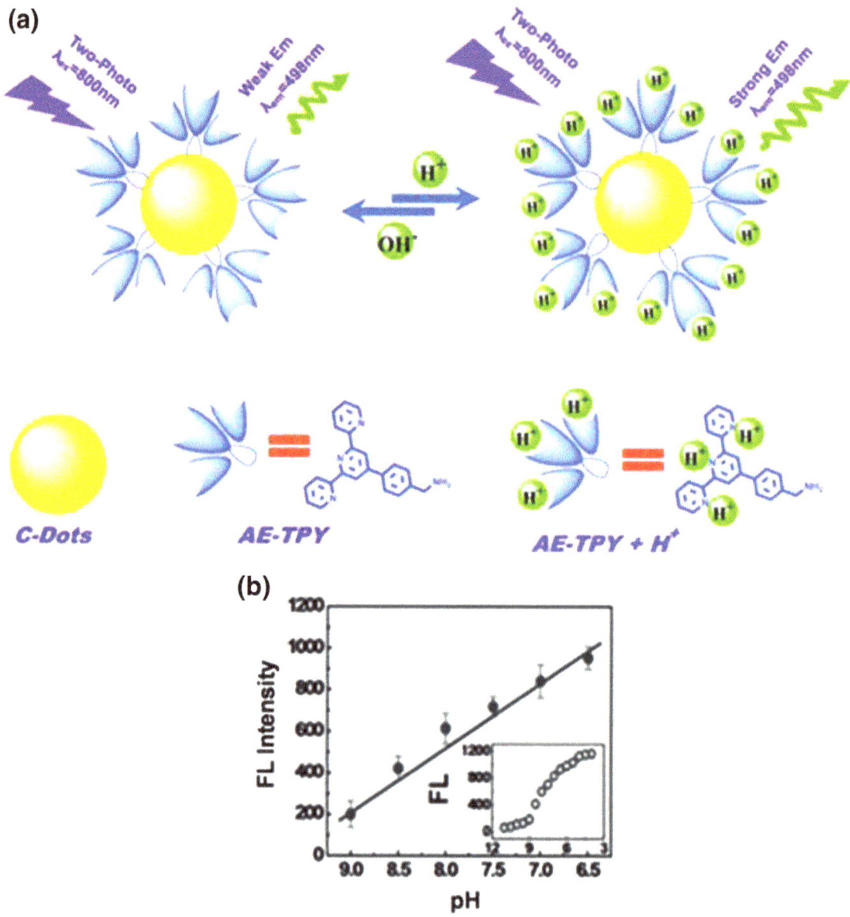

Fig. 6.1 pH sensing by surface-functionalized carbon-dots. **a** Structure of the Carbon-Dots: a pH-sensitive receptor molecule, aminomethylphenyl-terpyridine (AE-TPY) is covalently attached to the Carbon-Dots surface. Proton adsorption onto the TPY units presumably results in modulation of the surface energy states of the Carbon-Dots, consequently affecting the intensity of fluorescence emission. **b** Dependence of the Carbon-Dots' fluorescence intensity upon solution pH. Shown is the physiological range, and complete range (inset). Excitation was at 498 nm. Reprinted from Kong et al. [1], with permission from John Wiley & Sons (2012)

Carbon-Dots' surface induced modulation of fluorescence *intensity* following adsorption/desorption of H^+ ions. The pH dependence was linear in the physiological pH range (6.5–8.5), making possible pH sensing in live cells and tissues.

Metal ion detection has likely been the most ubiquitous sensing application carried out with Carbon-Dots, both in biological and chemical contexts. This is partly due to the affinity and coordination of varied metal ions with *aromatic* residues, which constitute prominent structural features of Carbon-Dots. A common consequence of metal binding to Carbon-Dots has been the occurrence of *luminescence quenching*, in some instances induced by extremely low metal ion concentrations, thus providing an effective vehicle for metal sensing. Moreover, *ionic selectivity* has been observed in many systems, further contributing to the use of Carbon-Dots as metal ion sensors.

The prominence of many metal ionic species in biological systems and biological processes has been a significant driving force for development of Carbon-Dot-based ion sensing applications. G. Shi and colleagues at East China Normal University have developed, for example, an electrochemical sensor for Cu^{2+} ions based upon specific capture of these divalent ions by functionalized Carbon-Dots [2]. As depicted in Fig. 6.2, the researchers combined two innovative schemes in the sensor concept. First, a receptor recognizing copper ions was attached onto the Carbon-Dots' surface, endowing them with ion specificity. Second, the modified Carbon-Dots were displayed upon the surfaces of the *electrodes* in an electrochemical cell, enabling efficient electron transport and occurrence of redox reactions upon their surface. Notably, the Carbon-Dots were *first* immobilized onto the electrode surface and then underwent the chemical reaction resulting in covalent display of the Cu^{2+} receptor upon their surface; the reaction did not adversely affect the integrity and photophysical properties of the Carbon-Dots. Interestingly, the Carbon-Dot-containing electrochemical sensor accomplished Cu^{2+} detection in rat brain lysates extracted through an "online" setup as depicted in Fig. 6.2a.

A different strategy for high sensitivity detection of extracellular cerebral Cu^{2+} is depicted in Fig. 6.3. The detection system, developed by Y. Lin and colleagues at Capital Normal University, China, was based upon *quenching* of the Carbon-Dots' fluorescence through induced aggregation of the carbon nanoparticles [4]. Specifically, as illustrated in Fig. 6.3a, the Carbon-Dots were coated with a silica layer displaying terminal amine residues. Intercalation of copper ions between the amine residues of adjacent C-cots gave rise to formation of extended Carbon-Dot networks and subsequent aggregation; this process was accompanied by *quenching* of the fluorescence emission of the Carbon-Dots, ascribed both to close proximity between the Carbon-Dots as well as energy transfer from the Carbon-Dots to the d-orbitals of the Cu^{2+} ions. The silica-coated Carbon-Dot system exhibited high selectivity and sensitivity to Cu^{2+} and sensor operation was demonstrated in actual brain microlysate dialysis (Fig. 6.3b).

Aggregation-induced quenching of Carbon-Dots' fluorescence has been quite a popular sensing mechanism due to the varied pathways in which aggregation/disaggregation of Carbon-Dots could be induced. Figure 6.4 depicts an example of

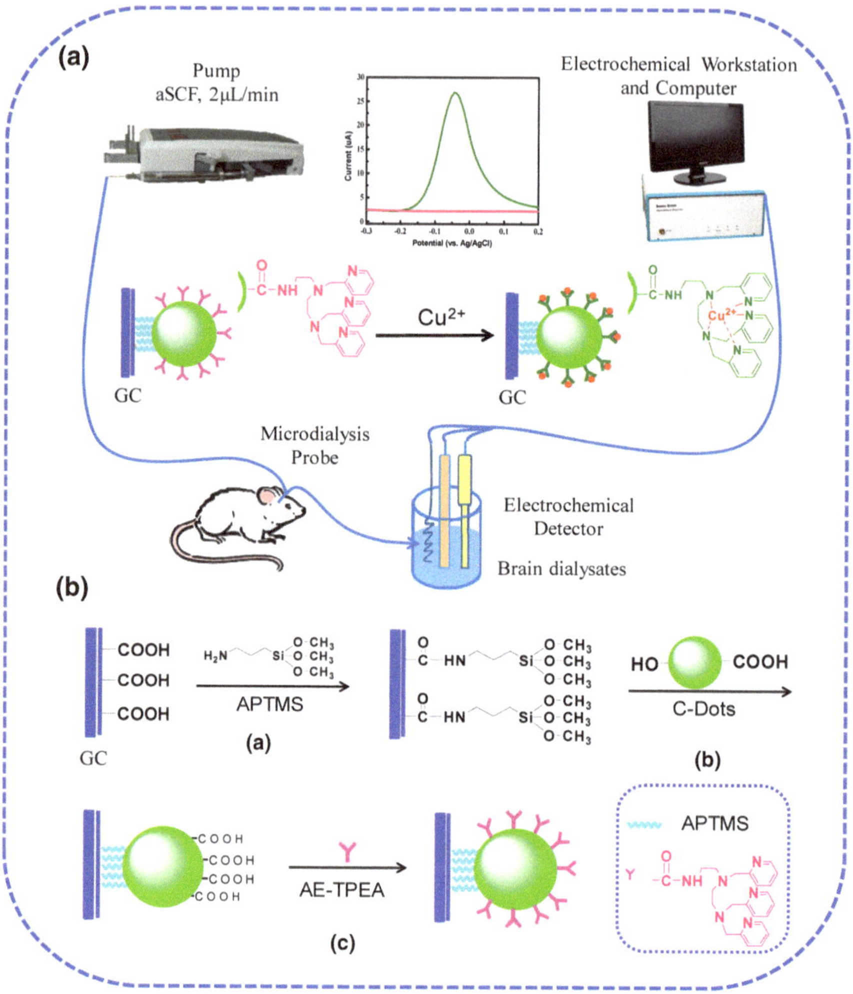

Fig. 6.2 Cu^{2+} sensing by carbon-dots functionalized with a copper-recognition ligand. **a** Overall experimental setup. The Carbon-Dots displaying Cu^{2+}-recognition ligands are covalently linked to the surface of the glassy carbon (GC) electrode in an electrochemical cell. The electrochemical detector allows continuous sensing of cerebral Cu^{2+} levels in rats. **b** Construction of the Carbon-Dot-containing electrodes. Note that functionalization of the Carbon-Dots is carried out after they are attached to the GC surface [3]

such a sensing strategy. The Carbon-Dots, designed by M. Yang and colleagues at Central South University, China, were covalently coupled to antibodies and aptamers (short oligonucleotides that are selective to target biomolecules) directed at *mucin-1*, a known tumor marker [5]. Accordingly, in the presence of mucin-1 the Carbon-Dots were cross-coupled to form large assemblies, resulting in quenching of the Carbon-Dots' fluorescence. The fluorescence modulation exhibited linear

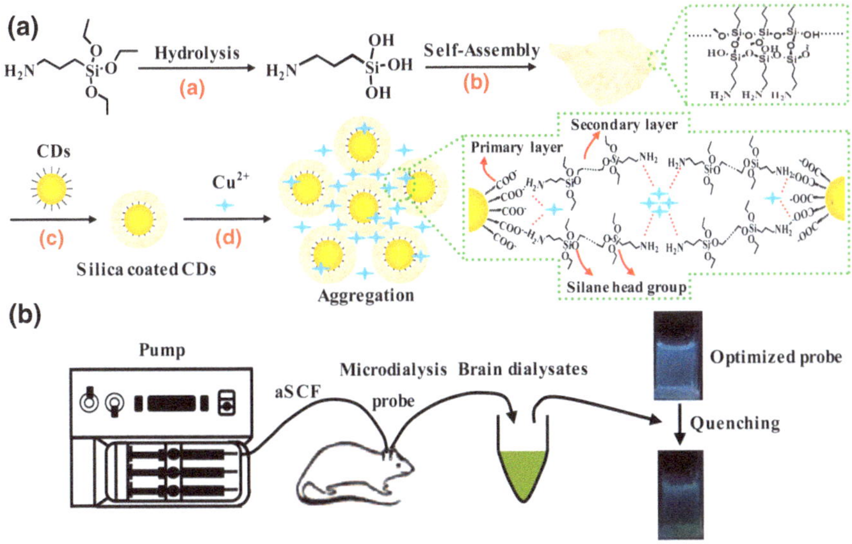

Fig. 6.3 Cu^{2+} sensing in physiological fluids with silica-coated carbon-dots. **a** Synthesis scheme. The Carbon-Dots are coated with a primary silica layer and a secondary silane layer. Intercalation of the calcium ions results in aggregation and subsequent quenching of the Carbon-Dots' fluorescence. **b** Practical application of the Carbon-Dot sensor for Cu^{2+} analysis in rat brain lysate. Reprinted with permission from Lin et al. [4]. Copyright (2015) American Chemical Society

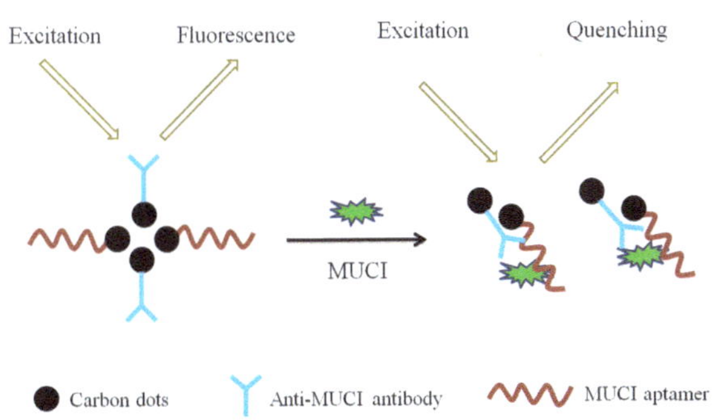

Fig. 6.4 Mucin-1 (MUC1)-induced quenching of carbon-dots' fluorescence. The Carbon-Dots are functionalized with MUC1 aptamers and anti-MUC1 antibodies. MUC1-induced Carbon-Dot aggregation results in fluorescence quenching Adapted from Ma et al. [5]

response to mucin-1 concentrations in the *nanomolar* range, underscoring the potential practicality of this sensing concept.

A composite sensor based upon Carbon-Dot aggregation/disaggregation and designed to detect the amino acid *cysteine* is illustrated in Fig. 6.5. The sensor system, developed by Y. Zhang and colleagues at Hunan Normal University, China, comprised of a core assembly of Au nanoparticles (NPs) surrounded by a shell of

Carbon-Dots [6]. The detection mechanism exploited the affinity of the thiol residues of cysteine to the Au NPs' surface. In the initial state, the fluorescence of the Carbon-Dots in the shell environment was quenched because of their close proximity. However due to Au NP/cysteine affinity, the cysteine residues "penetrated" through the Carbon-Dot shell and bound to the Au NP surface, thereby dispersing the Carbon-Dots in solution, resulting in increased fluorescence. Notably, this scheme could distinguish between cysteine and other thiol-containing molecules such as homocysteine and glutathione (GSH); the latter two molecules exhibit lesser affinity to Au NPs and thus did not displace the Carbon-Dots (Fig. 6.5). Notably, the composite sensor in Fig. 6.5 provided signal modalities other than Carbon-Dots' fluorescence: cysteine-induced disaggregation of the *gold NPs* induced the well-known colorimetric (i.e., plasmon) shift from red/purple to blue.

Several publications described utilization of Carbon-Dots for *functional sensing* of biomolecules, particularly proteins and enzymes. Z. Qian and colleagues at Zhejiang Normal University, China, reported conjugation of Carbon-Dots with *dopamine* and application of the dopamine-modified Carbon-Dots for sensing of enzymatic activity of *tyrosinase* (*TYR*) [7]. TYR is involved in important cellular redox processes; specifically, this enzyme is pivotal in biosynthesis of *melanin*—a prominent skin-protective pigment. Detection of TYR activity and the effect of potential enzyme inhibitors are thus highly sought. The Carbon-Dot-based TYR detection mechanism, shown in Fig. 6.6, relied on attachment of dopamine residues onto the carbon nanoparticles. When the enzyme was present, oxidation of

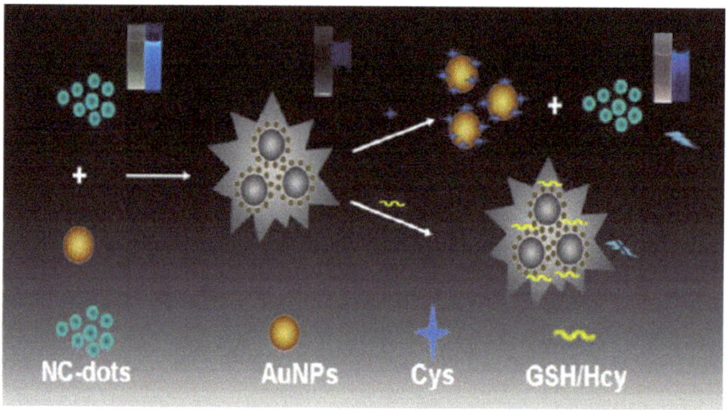

Fig. 6.5 Carbon-dot/gold composites for cysteine detection through fluorescence quenching. A scheme depicting the assembly method and cysteine detection approach: The Carbon-Dots (denoted "nanocarbon-dots", NC-dots) are mixed with gold nanoparticles (AuNPs) and initially form "shells" around the AuNPs; the Carbon-Dots fluorescence is quenched due to close proximity of the Carbon-Dots (*middle*). Cysteine affinity to the Au NP disintegrates the Carbon-Dot shells, and the Carbon-Dots' fluorescence is resurrected due to their dispersion in the aqueous solution (*top right*). Thiol-containing molecules such as GSH or homocysteine (Hcy) do not displace the Carbon-Dot shells as they exhibit less affinity to the AuNP surface and the Carbon-Dots' fluorescence is still quenched (*bottom right*). Reprinted with permission from Deng et al. [6]. Copyright (2015) American Chemical Society

dopamine occurred, resulting in its transformation to dopaquinone (Fig. 6.6a). Consequent photo-induced electron transfer (PET) from the Carbon-Dot to the quinone moieties resulted in *quenching* of the Carbon-Dot fluorescence. However, when *inhibitors* of TYR were added, the Carbon-Dot-displayed dopamine was *not* oxidized by the enzyme, resurrecting the Carbon-Dot fluorescence. As shown in Fig. 6.6b, the TYR sensing assay was highly selective, reporting on the presence of the enzyme in physiological solutions containing other biological molecules.

Forster resonance energy transfer (FRET) has been employed as a useful mechanism for Carbon-Dot-based biosensing. Carbon-Dots, in fact, could serve as either fluorescent *donors* or *acceptors* in FRET platforms. Figure 6.7 presents a

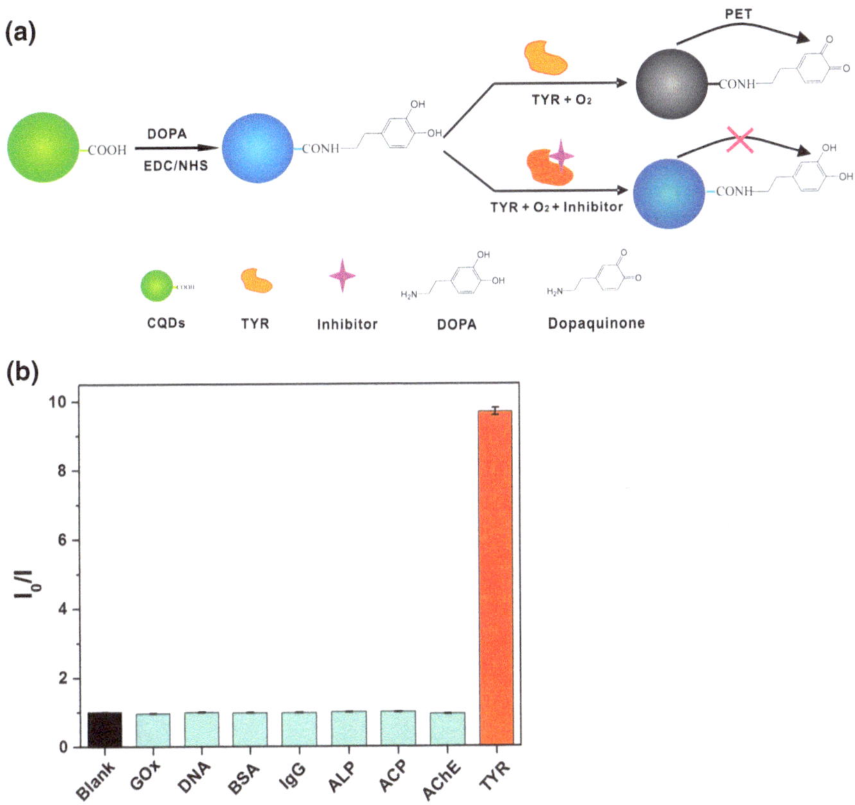

Fig. 6.6 Dopamine-functionalized carbon-dots for *tyrosinase* (*TYR*) sensing. **a** Synthesis scheme showing coupling of dopamine to the Carbon-Dots. In the presence of the TYR enzyme, dopamine is oxidized to dopaquinone and a photo-induced energy transfer (PET) process quenches the Carbon-Dot's fluorescence. Co-addition of TYR inhibitors blocks dopamine oxidation and the fluorescence is retained. **b** Bar diagram demonstrating selectivity of the sensing system to TYR; significantly *lower* quenching of the Carbon-Dots' fluorescence was recorded in the presence of other enzymes and biomolecules. Reprinted with permission from Chai et al. [7]. Copyright (2015) American Chemical Society

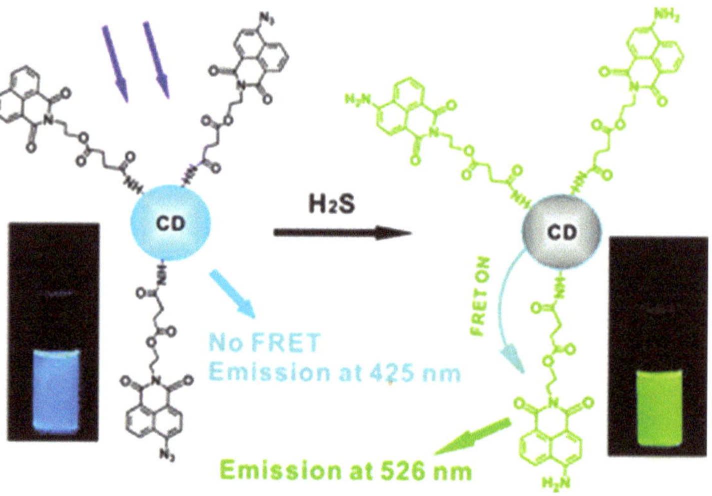

Fig. 6.7 H_2S sensing through FRET in functionalized carbon-dots. No FRET between the Carbon-Dot core and covalently attached naphthalimide azide residues occurs (*left*). However, H_2S reduces the azide to amine, generating naphthalimide amine which is an energy acceptor in FRET experiments. *Photographs* of the respective Carbon-Dot solutions are shown—*blue* luminescence of the Carbon-Dots is recorded without H_2S (*left*), while *green* luminescence of naphthalimide amine is apparent after H_2S-induced reduction. Reproduced from Yu et al. [8], with permission of the Royal Society of Chemistry

FRET-based biosensor in which the Carbon-Dots served as fluorescence donors towards detection of *hydrogen sulfide*, a biological signaling molecule and a putative marker in several diseases [8]. In this study, S. Wu and colleagues at South China University of Technology synthesized Carbon-Dots that were covalently linked to a naphthalimide azide derivative. Initially, no FRET occurred from the Carbon-Dot core to naphthalimide azide (Fig. 6.7). However, naphthalimide *amine*—a fluorescence *acceptor* moiety—was formed following *reduction* by H_2S, facilitating efficient FRET from the Carbon-Dot acting as an energy donor. Importantly, FRET made possible high sensitivity H_2S detection (i.e., low detection threshold), as well as relatively large dynamic range which covered physiological concentrations of the molecule. Indeed, this study exemplifies the advantages of FRET as a sensing strategy, accomplishing in many instances superior sensing properties. It should be mentioned, however, that applications of FRET often require careful experimental design and occasionally complex synthetic routes which might be detrimental to broad-based biosensing applications.

Interesting applications of Carbon-Dots as sensors for ***radicals*** have been developed by several research groups. Radical species are associated with many diseases and pathological conditions, including cardiovascular diseases, neurological pathologies, and others, and accordingly their study both in vitro and in vivo is of significant interest. The thrust of Carbon-Dot-based radical sensing applications has been the effect of radicals upon the *fluorescence intensity* emitted by the

Carbon-Dots, induced by electron transfer processes. Pronounced fluorescence *quenching*, for example, was reported by X. Guo and colleagues at Xiamen University, China, when Carbon-Dots were brought into contact with nitroxide radicals ($NO^{\cdot}$), presumably through electron transfer from the dots to the radical molecules [9]. This feature could be employed for detection of potential antioxidants—as such compounds *prevent* radical formation or eliminate them in physiological conditions, thereby increasing the Carbon-Dots' fluorescence.

Other Carbon-Dot-based radical sensors have employed *ratiometric* fluorescence sensing. Ratiometric sensors utilize the *ratio* between the fluorescence intensities at *two* wavelengths as a yardstick for the presence and concentration of a target analyte. As such, ratiometric measurements are considered more accurate than simple emission-based sensing schemes since the ratio of the fluorescent intensities at two wavelengths is independent of probe concentrations and possible inhomogeneous distributions, and additionally circumvents technical issues such as light source fluctuations. Figure 6.8 presents a Carbon-Dot system designed to detect *superoxide* (O_2^-), the reaction product of a single-electron reduction of molecular oxygen. Superoxide is a precursor for reactive oxygen species (ROS), considered toxic and disease-inducing substances, and as such its detection in physiological environments is highly sought. The sensing configuration in Fig. 6.8, developed by Y. Tian and colleagues at East China Normal University, comprised of Carbon-Dots to which a fluorescent dye recognizing O_2^- was covalently attached [10]. While the fluorescence intensity emitted by the superoxide-binding dye was directly dependent upon superoxide concentration, the fluorescence of the *Carbon-Dots* (in a different wavelength) was *independent* of the superoxide presence; accordingly, the ratio between the two fluorescence emissions was a sensitive marker for superoxide concentration. Figure 6.8b demonstrates application of the Carbon-Dot conjugate for *ratiometric imaging* of cells. Specifically, the confocal fluorescence microscopy images in Fig. 6.8b clearly reveal elevated intracellular superoxide levels, induced by lipopolysaccharides (LPS), known stimulators of ROS production.

One of the most powerful features of Carbon-Dots as sensing vehicles is the feasibility of bestowing the Carbon-Dots with molecular recognition elements (e.g., selective binding) simply through the use of synthetic building blocks containing such binding units. This approach is conceptually and practically easier to carry out compared to post-synthesis chemical coupling of recognition elements, which has been commonly employed in numerous sensor systems. Figure 6.9 depicts a *glucose sensing* platform comprising Carbon-Dots that were prepared through hydrothermal condensation of phenylboronic acid as the carbonaceous precursor [11]. In that study, Y. Xia and P. Shen at Anhui Normal University, China, demonstrated that the fluorescence of the boronic acid-generated Carbon-Dots could be employed for high sensitivity detection of sugar, even in physiological blood environments. Specifically, glucose molecules induced *quenching* of the Carbon-Dots' fluorescence, presumably through chelating and aggregating Carbon-Dots via the boronic acid receptors upon the dot surface.

Surface modification of Carbon-Dots has been employed as the means for ***bacterial sensing***. Display of functional residues on Carbon-Dots allows their

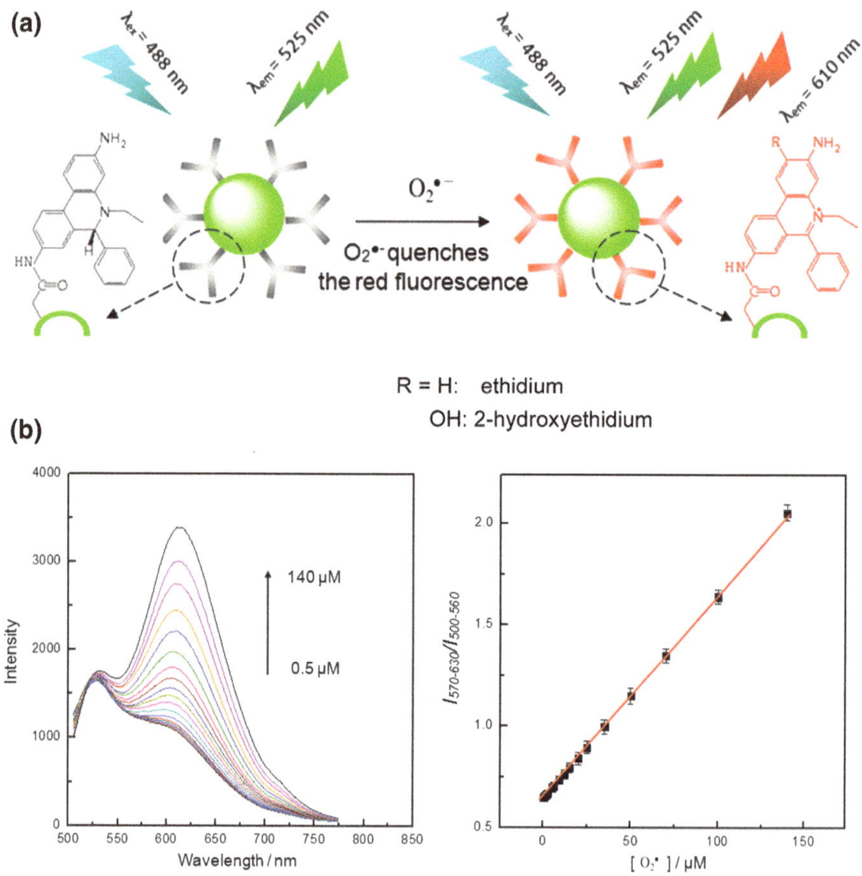

Fig. 6.8 Ratiometric fluorescence sensing of a superoxide radical by functionalized carbon-dots. **a** The sensing concept: a superoxide-recognition ligand is covalently linked to the Carbon-Dots. The fluorescence of the ligand is directly dependent upon superoxide binding. **b** Results of the ratiometric experiment: the Carbon-Dot fluorescence emission peak at 530 nm (exc. 488 nm) is not affected by superoxide presence, while the fluorescence emission of the superoxide-binding ligand at 620 nm becomes more intense upon increasing superoxide concentrations. **c** Ratiometric cell imaging (shown are the topographic maps of the ratio between fluorescence intensity at 570–630 nm, and intensity at 500–560 nm. Excitations: 488 nm). *a* Control HeLa cells (*b* Bright field image). *c–d* HeLa cells treated with LPS for 40 and 60 min, respectively. Note the higher abundance of red areas within the cells, reflecting higher fluorescence ratios. Reprinted with permission from Gao et al. [10]. Copyright (2014) American Chemical Society

targeting and binding onto bacterial cells. *Bacterial specificity*, however, is considered a major challenge in such sensing applications (as is the case also in many other bacterial sensing methods). Figure 6.10 depicts a bacterial sensing scheme based upon *amphiphilic Carbon-Dots*, i.e., Carbon-Dots onto which long hydrocarbon chains were attached, bestowing the particles with high affinity to bacterial membranes [12]. Indeed, simple addition of the amphiphilic Carbon-Dots

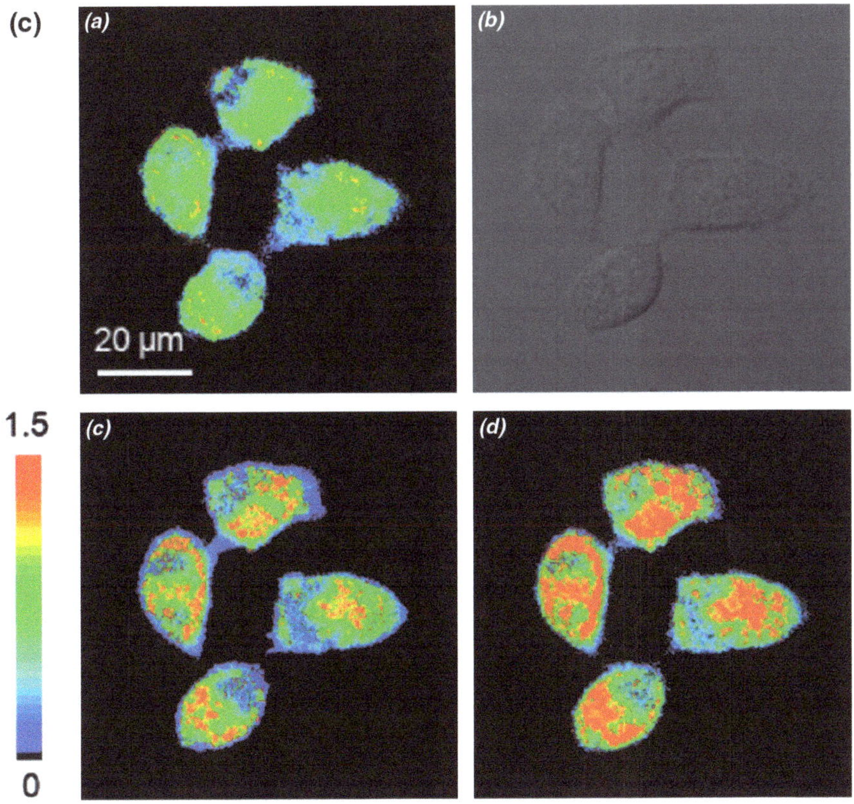

Fig. 6.8 (continued)

to the bacterial growth media resulted in adsorption of the dots to the surface of the bacterial cells, making them highly fluorescent (Fig. 6.10). Notably, the distinct membrane compositions of different bacterial species made possible bacterial *discrimination* based upon changes in the fluorescence emission shift and intensity of the Carbon-Dots. This potentially useful feature relies upon the sensitivity of Carbon-Dots' luminescence towards the molecular environment of the particles, reflected in the different molecular membrane compositions of bacterial families and strains.

A strategy designed to achieve specific fluorescence bacterial labeling through the use of *aptamer coupling* onto Carbon-Dots is depicted in Fig. 6.11. Aptamers are short oligonucleotides capable of selective, high affinity binding to diverse molecular targets. In the study outlined in Fig. 6.11, Y. Xu and colleagues at Chongking University, China, covalently attached Carbon-Dots and aptamers targeted to surface proteins of the bacterium *Salmonella typhimurium* [13]. The Carbon-Dot-aptamer composite particles displayed good fluorescence properties, particularly high brightness, and could be employed for selective labeling of *S. typhimurium* and quantification of bacterial cells. Beside *specificity* to bacteria type,

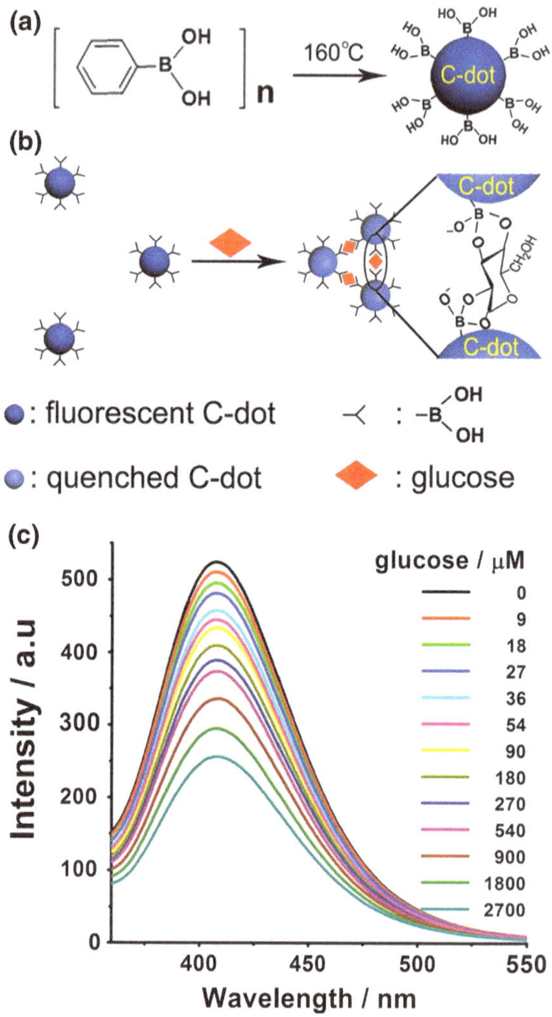

Fig. 6.9 Glucose sensing by carbon-dots prepared from boronic acid as the carbon source. **a** Synthesis of the Carbon-Dots through a one-step hydrothermal treatment of boronic acid; **b** glucose sensing concept: glucose molecules induce intercalation of Carbon-Dots via the boronic acid receptor, consequently inducing quenching of Carbon-Dots' fluorescence; **c** progressive quenching of Carbon-Dots' fluorescence upon increasing glucose concentrations. Reprinted with permission from Shen and Xia [11]. Copyright (2014) American Chemical Society

this Carbon-Dot labeling technique exhibited very low detection threshold—as low as ~1000 cells/mL—an excellent value that might enable practical implementation of this bacterial sensing method.

6.2 Carbon-Dots in Chemical Sensing

Mercury ion (Hg^{2+}) detection has been among the most widely reported chemical sensing applications employing Carbon-Dots. Mercury ions are toxic, and since they might be present in many food products, development of sensors for these ions has attracted significant R&D activity. Varied configurations of Carbon-Dot-based

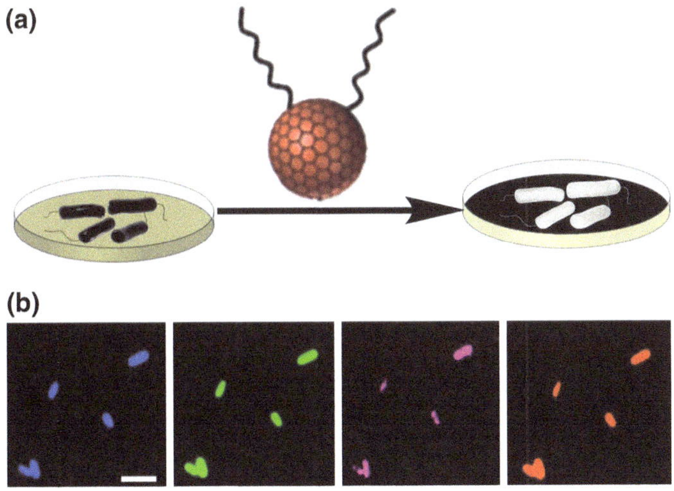

Fig. 6.10 Bacterial staining with amphiphilic carbon-dots. **a** Bacteria are incubated with Carbon-Dots displaying long hydrocarbon chains upon their surface (i.e. amphiphilic Carbon-Dots). The amphiphilic Carbon-Dots adsorb onto the bacterial cell membrane, enabling visualization in fluorescence microscopy (**b**). The colors reflect different excitation/emission pairs. Adapted from Nandi et al. [12]

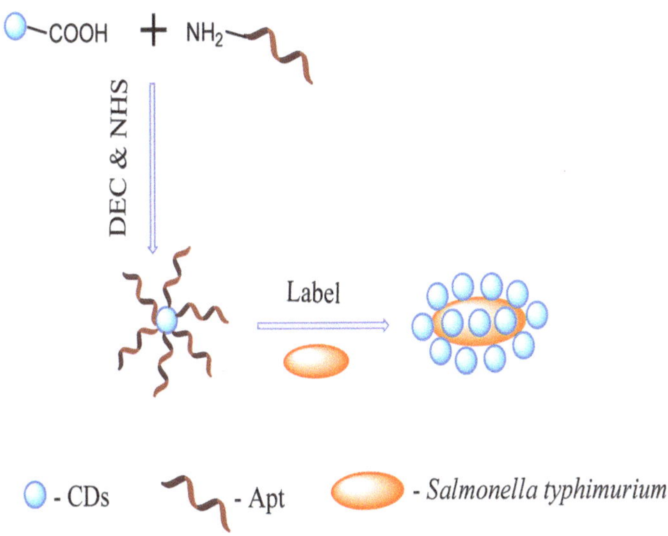

Fig. 6.11 Bacterial labeling through the use of Carbon-Dot-aptamers. Aptamers directed against bacterial cell proteins are covalently attached to the Carbon-Dots' surface and provide detection specificity. Reproduced from Wang et al. [13], with permission of the Royal Society of Chemistry

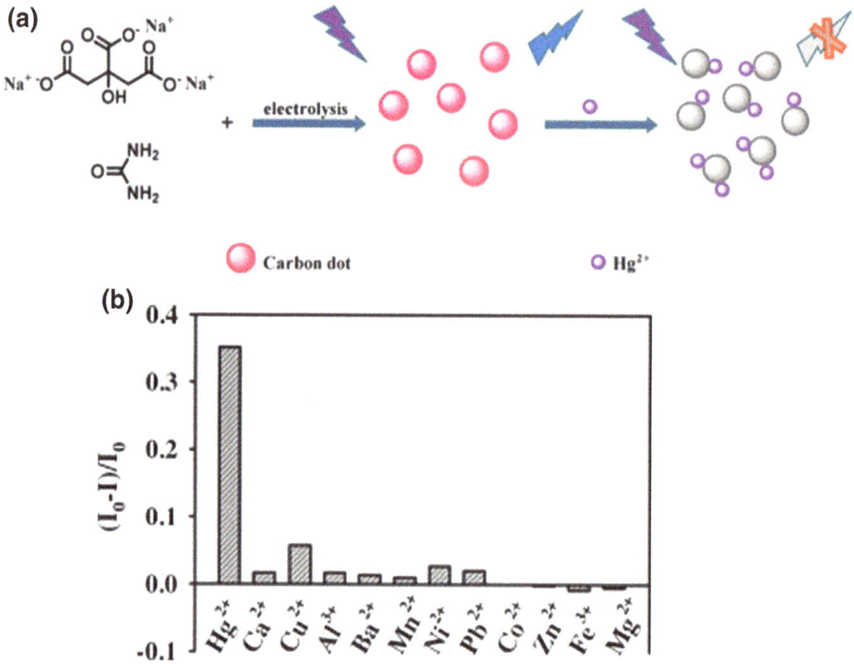

Fig. 6.12 Hg^{2+} sensing by carbon-dots. **a** Scheme showing the synthesis of Carbon-Dots through electrolysis of sodium citrate/urea mixture, and quenching of the Carbon-Dots' fluorescence following binding to Hg^{2+} ions; **b** selectivity of the assay—significantly more pronounced fluorescence quenching induced by mercury ions as compared to other metal ions. Reprinted from Hou et al. [14], Copyright (2015), with permission from Elsevier

Hg^{2+} sensors have been reported, designed to offer means for practical detection of the ions, particularly in field settings. Figure 6.12 depicts a representative Hg^{2+} sensing application accomplished with Carbon-Dots. The Carbon-Dots were prepared by Y. Zhang and colleagues at Hunan Normal University, China, from simple carbonaceous building blocks (sodium citrate and urea) [14]. Importantly, significant *quenching* of the Carbon-Dots' fluorescence was observed following addition of small Hg^{2+} concentrations. The quenching mechanism has not been elucidated, and is believed to occur through non-radiative energy transfer processes induced by chelation of the mercury ions by carboxylic residues upon the Carbon-Dots' surface. Interestingly, Carbon-Dot-based sensing of mercury was highly selective (Fig. 6.12b), likely reflecting the high chelating affinity of the Hg^{2+} ions toward the Carbon-Dots.

While *electrostatic attraction* has been widely utilized as the underlying mechanism for Hg^{2+} detection using Carbon-Dots (e.g., Fig. 6.12), a different model has been proposed in a Carbon-Dot system developed for sensing *methylmercury* (CH_3Hg^+), a potent neurotoxin associated with fish consumption. Methylmercury is more hydrophobic than the bare ion and is thus capable of crossing physiological lipid barriers. Accordingly, C. Bendicho and colleagues at the Universidad de Vigo, Spain, synthesized Carbon-Dots that were coated with a water-repulsing polyethyleneglycol (PEG) layer [3] (Fig. 6.13). Unlike many of the conventional hydrophilic Carbon-Dots, the PEG-coated dots *could not* be used for Hg^{2+} sensing since the ions did not interact with the surface of the nanoparticles. However, in very basic solutions (in which the assay was carried out) CH_3Hg^+ transformed to CH_3HgOH and/or $Hg(OH)_2$; both species are hydrophobic and could transport through the PEG coating, consequently quenching the Carbon-Dots' luminescence. An important feature of the methylmercury sensor assay depicted in Fig. 6.13 was the fact that both synthesis and analyte detection were carried out *simultaneously*; the C-cots were synthesized by ultrasonication (US) of the reagent mixture and in parallel fluorescence emission was monitored to account for the presence of methylmercury.

Figure 6.14 illustrates application of Carbon-Dots prepared through hydrothermal treatment of citric acid and ethylenediamine for selective sensing of Fe^{3+} ions. The experiments, carried out by B. Yang and colleagues at Jilin University, China, vividly demonstrate the dramatic quenching of the Carbon-Dots' fluorescence signal by ferric ions [15]. The selectivity of the Carbon-Dots apparent in Fig. 6.14b was ascribed to the strong multivalent coordination between Fe^{3+} and hydroxyl-phenolic residues presumably abundant upon the Carbon-Dots' surface. Similar to the Hg^{2+} sensing system discussed above (e.g., Fig. 6.12), such surface binding would allow rapid transfer of the excitons from the phenol moieties to

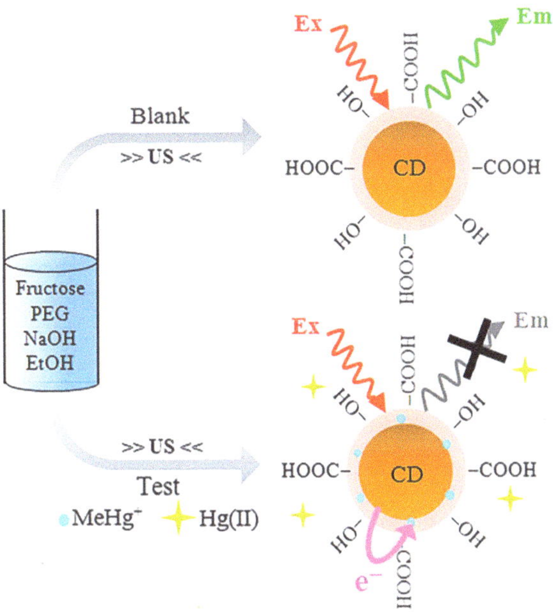

Fig. 6.13 Detection of methylmercury ($MeHg^+$) by carbon-dots coated with polyethyleneglycol (PEG). The one-step synthesis/detection scheme is illustrated. The Carbon-Dots are synthesized from fructose and PEG in basic conditions using ultrasonication (US). The $MeHg^+$ ions can penetrate through the PEG layer, quenching the Carbon-Dots' fluorescence (bottom). Reprinted with permission from Costas-Mora et al. [3]. Copyright (2014) American Chemical Society

d-orbitals of the iron ion in non-radiative processes, thereby quenching luminescence emission.

As discussed above, metal ion-induced quenching of Carbon-Dots' emission was recorded for a variety of ions and Carbon-Dot systems. Some Carbon-Dots have been shown to facilitate *multimodal* detection of metal ions, i.e., probing the presence of *different* ions using the *same* Carbon-Dots. Figure 6.15 presents an example of this concept. The Carbon-Dot platform, developed by H. Lin and colleagues at the Ningbo Institute of Materials Technology and Engineering (NIMTE), China, was synthesized through hydrothermal treatment of citric acid and formamide, and could produce intense fluorescence emission peaks in a broad range of excitation wavelengths (which is different than most Carbon-Dots reported so far, for which high emission peaks appear in relatively narrow excitation wavelengths windows) [16]. As demonstrated in Fig. 6.15, specific emission peaks enabled the researchers to distinguish among different metal ions as they exhibited different degrees of quenching in distinct excitation/emission wavelengths. For example, while Hg^{2+} gave rise to similar quenching in three different excitation wavelength, Cu^{2+} induced significantly *lower* quenching upon excitation at 360 nm in comparison with the two other wavelengths examined (Fig. 6.15a). A statistical evaluation using the principle component analysis (PCA) methodology revealed

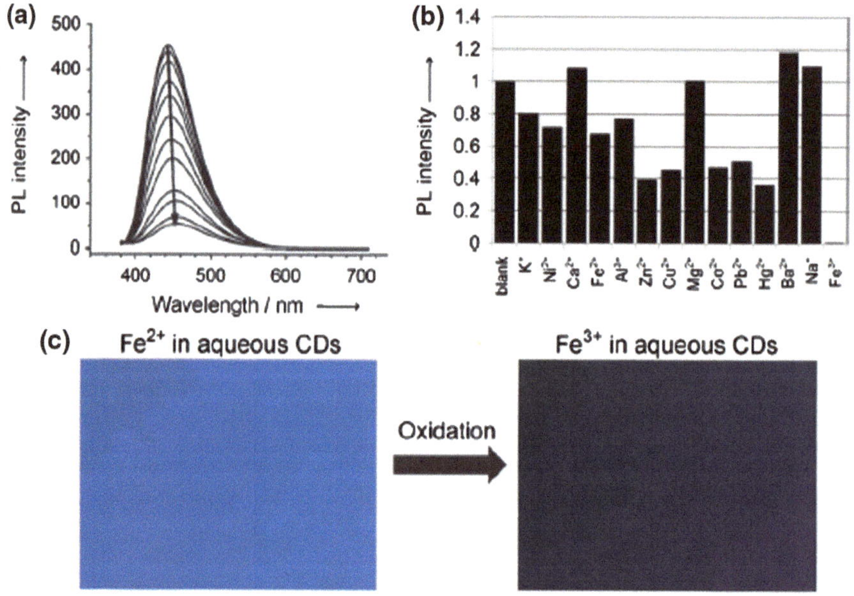

Fig. 6.14 Fe^{3+} sensing by carbon-dots. The Carbon-Dots were prepared from citric acid and ethylenediamine as carbon sources. **a** Quenching of the Carbon-Dots' fluorescence by progressive addition of Fe^{3+} ions. **b** Selectivity of the Carbon-Dot assay. Significantly more pronounced quenching observed in the presence of ferric ions. **c** Fluorescence microscopy images depicting the quenching effect induced by Fe^{2+} oxidation. Reprinted from Zhu et al. [15], with permission from John Wiley & Sons (2013)

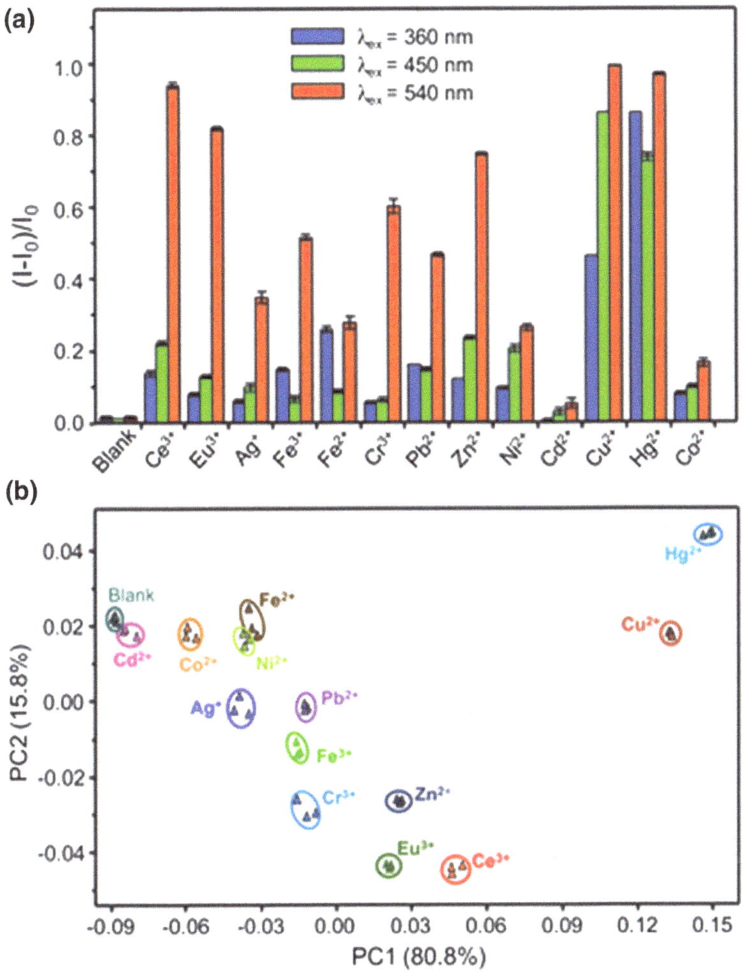

Fig. 6.15 Metal ion discrimination by "multimodal" sensing using carbon-dots. Distinct combinations of emission signals acquired upon excitation in three different wavelengths were observed. **a** Bar diagram depicting the extent of fluorescence emission *quenching* by the indicated metal ions following excitations in three distinct wavelengths. **b** Statistical evaluation of the emission results using principle component analysis (PCA). The groupings apparent in the two-dimensional graph indicate statistically-significant differences in fluorescence quenching induced by the ions in the three excitation wavelengths. Reprinted from Pan et al. [16], with permission from John Wiley & Sons (2015)

distinct signals associated with each ion upon the use of the three excitation wavelengths (Fig. 6.15b). The variations of fluorescence quenching were ascribed to distinct binding modes of the ions onto the Carbon-Dots' surface, thereby affecting differently the surface energy states and corresponding fluorescence profiles.

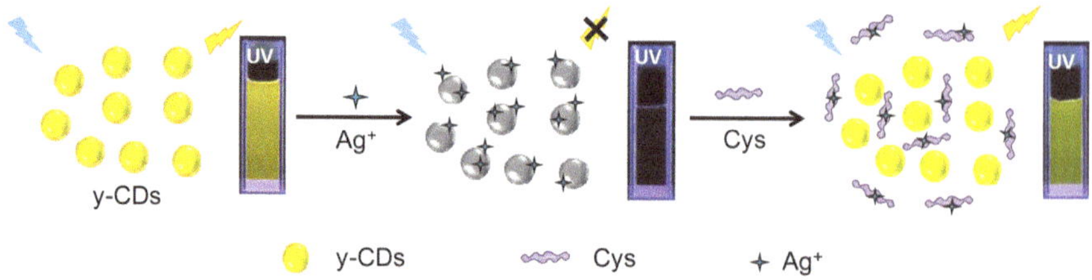

Fig. 6.16 Bifunctional sensing with yellow carbon-dots. Quenching of the Carbon-Dots' fluorescence recorded in the presence of Ag^+ ions. Addition of cysteine, however, induces "de-quenching" as the cysteine molecules bind to the Ag^+ ions. The Carbon-Dots can be employed accordingly for sensing of both Ag^+ and cysteine. Reprinted with permission from Jiang et al. [17]. Copyright (2015) American Chemical Society

An interesting Carbon-Dot-enabled *bifunctional* sensing application was recently reported by the same researchers [17]. As depicted in Fig. 6.16, the yellow-emissive Carbon-Dots displayed both sensitivity to Ag^+ ions which induced *quenching* of the Carbon-Dots' fluorescence, as well as *de-quenching* through binding of the amino acid *cysteine*. Resurrection of the fluorescence emission was explained by H. Lin and colleagues through binding of the *thiol* moieties of cysteine to the silver ions, thus blocking disruption of the Carbon-Dots' surface states. On a practical level, the Ag^+-bound yellow CDs were highly selective towards cysteine; other amino acids had negligible de-quenching effect.

While the applications described above employed Carbon-Dots for detection of target analytes in *solution phases,* some Carbon-Dot-based platforms were

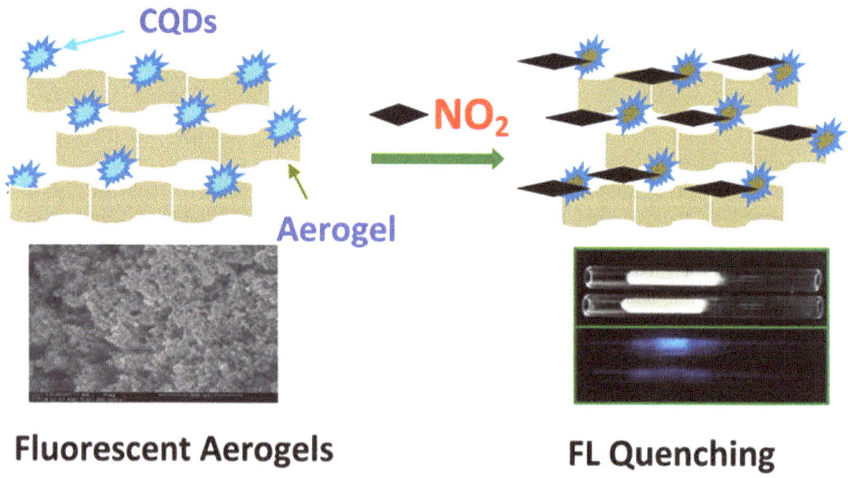

Fig. 6.17 NO_2 gas sensing by aerogel/carbon dot composite matrix. The Carbon-Dots are embedded within the aerogel pores; adsorption of NO_2 molecules within the aerogel induces quenching of the Carbon-Dots' fluorescence. Reprinted with permission from Wang et al. [18]. Copyright (2013) American Chemical Society

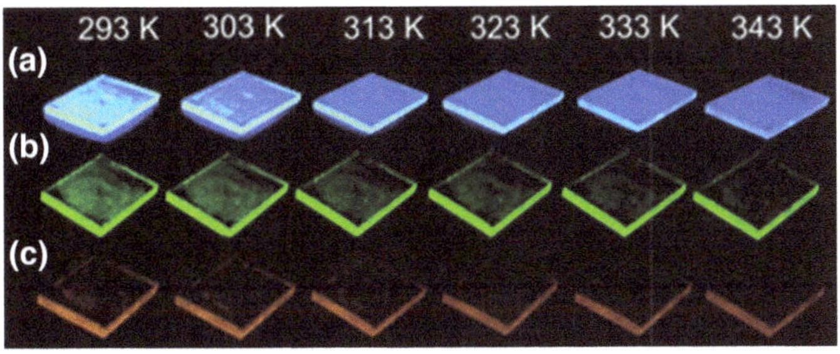

Fig. 6.18 Temperature sensitive silicon-carbon dot composite. The luminescence intensity is reversibly dependent upon temperature; the effect if apparent in different excitation wavelengths. Reproduced from Chen et al. [19], with permission of the Royal Society of Chemistry

developed for ***gas sensing***. Y. Chi and colleagues at Fuzhou University, China, synthesized an aerogel/Carbon-Dot matrix and demonstrated its application as a gas sensor [18]. Aerogels are low density porous materials that display substantial surface areas available for gas adsorption. The system developed by the researchers comprised of a silica-based aerogel encapsulating Carbon-Dots. The Carbon-Dots retained their high fluorescence emission even upon drying the Carbon-Dot/aerogel conjugate—a notable observation since the luminescence of Carbon-Dots has been shown to significantly diminish in many instances upon immobilization due to self-quenching. Particularly interesting was the *luminescence quenching* of the Carbon-Dots' fluorescence upon adsorption of NO_2 gas (Fig. 6.17). This phenomenon was ascribed to interactions between electron-rich amine moieties at the Carbon-Dots' surfaces (as the Carbon-Dots were synthesized from amine-displaying polymers) and the electron-withdrawing nitrogen dioxide gas molecules.

Carbon-Dots in varied configurations have been used as *temperature sensors*. An example for such a system is depicted in Fig. 6.18 In this study, H.T. Chang and colleagues at National Taiwan University prepared Carbon-Dots coupled to *silicon* through hydrothermal synthesis using organosilanes as precursor compounds [19]. Interestingly, the *luminescence intensities* of the Si-Carbon-Dots exhibited reversible modulation by the solution temperature. The mechanism responsible for this phenomenon has not been determined yet; presumably, the fluorescence changes were related to temperature-induced "energy traps" upon the Carbon-Dots' surface, possibly giving rise to energy transfer and quenching. The researchers have also shown that the presence of silicon enabled formation of stable Si-Carbon-Dot films on glass, held together through Si-O-Si bond formation.

Another interesting application of Carbon-Dots as "nano-thermometers" has been demonstrated by Z. Kang and colleagues at Soochow University, China [20]. The researchers observed that the fluorescence of nitrogen-doped Carbon-Dots exhibited distinct and linear dependence upon the solution temperature (Fig. 6.19). Temperature dependence of the fluorescence signal was reversible and the fluorescence signals could distinguish temperature increase/decrease up to fractions of

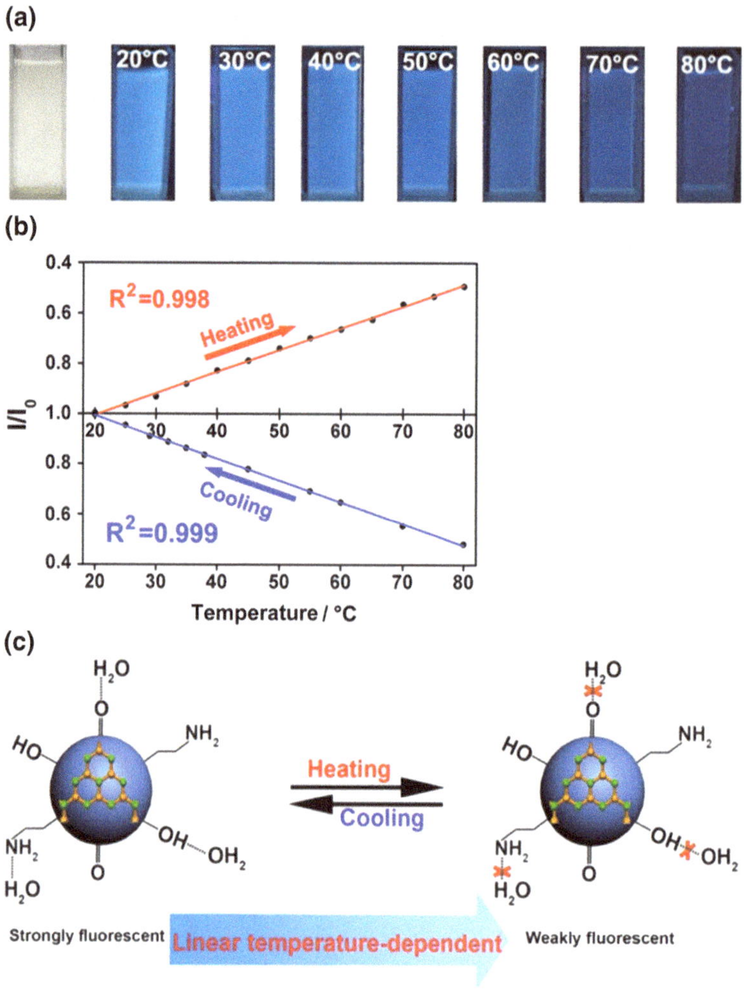

Fig. 6.19 Carbon dot nano-thermometer. **a** Temperature-dependent change in Carbon-Dots' fluorescence intensity; **b** linear temperature dependence of fluorescence increase/decrease; **c** proposed mechanism for temperature modulation of fluorescence intensity. Temperature increase breaks hydrogen bonding upon the Carbon-Dot surface thereby disrupting surface energy states. Reprinted with permission from Yang et al. [20]. Copyright (2015) American Chemical Society

degrees, pointing to possible practical applications of this Carbon-Dot system as a sensitive temperature sensor. The model proposed by the researchers for the temperature dependence pointed to disruption of hydrogen bonding between water molecules and oxygen- and/or amine residues upon the Carbon-Dots' surface (Fig. 4.19c). The putative surface transformations might modulate the surface-associated energy levels leading to different luminescence emission profiles.

References

1. B. Kong, A. Zhu, C. Ding, X. Zhao, B. Li, Y. Tian, Carbon dot-based inorganic–organic nanosystem for two-photon imaging and biosensing of pH variation in living cells and tissues. Adv. Mater. **24**, 5844–5848 (2012)
2. X. Shao, H. Gu, Z. Wang, X. Chai, Y. Tian, G. Shi, Anal. Chem. **85**, 418–425 (2013)
3. I. Costas-Mora, V. Romero, I. Lavilla, C. Bendicho, Anal. Chem. **86**, 4536–4543 (2014)
4. Y. Lin, C. Wang, L. Li, H. Wang, K. Liu, K. Wang, B. Li, ACS Appl. Mater. Interfaces **7**, 27262–27270 (2015)
5. N. Ma, W. Jiang, T. Li, Z. Zhang, H. Qi, M. Yang, Microchim. Acta **182**, 443–447 (2014)
6. J. Deng, Q. Lu, Y. Hou, M. Liu, H. Li, Y. Zhang, S. Yao, Anal. Chem. **87**, 2195–2203 (2015)
7. L. Chai, J. Zhou, H. Feng, C. Tang, Y. Huang, Z. Qian, ACS Appl. Mater. Interfaces **7**, 23564–23574 (2015)
8. C. Yu, X. Li, F. Zeng, F. Zheng, S. Wu, Chem. Commun. **49**, 403–405 (2013)
9. F. Lin, D. Pei, W. He, Z. Huang, Y. Huang, X. Guo, J. Mater. Chem. **22**(2012), 11801–11807 (1807)
10. X. Gao, C. Ding, A. Zhu, Y. Tian, Anal. Chem. **86**, 7071–7078 (2014)
11. P. Shen, Y. Xia, Anal. Chem. **86**, 5323–5329 (2014)
12. S. Nandi, M. Ritenberg, R. Jelinek, Analyst **140**, 4232–4237 (2015)
13. R. Wang, Y. Xu, T. Zhang, Y. Jiang, Anal. Methods **7**, 1701–1706 (2015)
14. Y. Hou, Q. Lu, J. Deng, H. Li, Y. Zhang, One-pot electrochemical synthesis of functionalized fluorescent carbon dots and their selective sensing for mercury ion. Anal. Chim. Acta **866**, 69–74 (2015)
15. S. Zhu, Q. Meng, L. Wang, J. Zhang, Y. Song, H. Jin, K. Zhang, H. Sun, H. Wang, B. Yang, Highly photoluminescent carbon dots for multicolor patterning, sensors, and bioimaging. Angew. Chem. Int. Ed. **52**, 3953–3957 (2013)
16. L. Pan, S. Sun, A. Zhang, K. Jiang, L. Zhang, C. Dong, Q. Huang, A. Wu, H. Lin, Truly fluorescent excitation-dependent carbon dots and their applications in multicolor cellular imaging and multidimensional sensing. Adv. Mater. **27**, 7782–7787 (2015)
17. K. Jiang, S. Sun, L. Zhang, Y. Wang, C. Cai, H. Lin, ACS Appl. Mater. Interfaces **7**, 23231–23238 (2015)
18. R. Wang, G. Li, Y. Dong, Y. Chi, G. Chen, Anal. Chem. **85**, 8065–8069 (2013)
19. P.-C. Chen, Y.-N. Chen, P.-C. Hsu, C.-C. Shih, H.-T. Chang, Chem. Commun. **49**, 1639–1641 (2013)
20. Y. Yang, W. Kong, H. Li, J. Liu, M. Yang, H. Huang, Y. Liu, Z. Wang, Z. Wang, T.-K. Sham, J. Zhong, C. Wang, Z. Liu, S.-T. Lee, Z. Kang, ACS Appl. Mater. Interfaces (2015)

Chapter 7
Materials Science Applications of Carbon-Dots

While numerous reports have described applications of Carbon-Dots in biological systems or their uses as luminescent sensors, there has been significant research activity towards exploiting Carbon-Dots in materials sciences in general, electro-optics, solar cells, and photonics, in particular. Similar to biological aspects of Carbon-Dot applications, the unique photoluminescence (PL) properties of these nanoparticles have been the main driving force for their implementation in diverse technologies and devices. Several examples of nonbiological materials applications of Carbon-Dots are discussed below.

Carbon-Dots have been proposed as core components in ***ink*** solutions. This application has been a natural outgrowth of the remarkable multicolor PL properties of Carbon-Dots, inexpensive reagents, and simple synthetic pathways. Figure 7.1 presents an elegant example of water-based ink comprising Carbon-Dots [1]. The Carbon-Dots were synthesized by L. Wang and colleagues at the Changchun Institute of Optics, Fine Mechanics and Physics, China, through a one-step microwave heating of a mixture of urea and citric acid (popular carbonaceous building blocks in Carbon-Dot synthesis). A recurring problem of Carbon-Dots when deposited and dried upon surfaces, impeding their use as potential ink materials, has been their *aggregation and quenching* upon drying. Notably, the Carbon-Dots synthesized by L. Wang and colleagues *did not* aggregate when deposited on paper and other organic fibrous substances, opening the way for varied ink-based applications. Interestingly, the Carbon-Dots could not be deposited upon *inorganic* substances. These observations are probably related to the affinity of the Carbon-Dot surface residues to specific functional units within the biological/organic substrate materials which blocked their aggregation (thus preventing fluorescence quenching). Figure 7.1 demonstrates intense fluorescence upon "printing" the Carbon-Dots upon conventional paper (Fig. 7.1a, b), and on cotton fiber but *not* nylon fiber (Fig. 7.1c–h).

Figure 7.2 depicts an intriguing application of Carbon-Dots in *inkjet printing* [2]. In that study, S. Chen and colleagues at Nanjing University of Technology, China, prepared new "ink" comprising Carbon-Dots and glycol, a solvent co-added to prevent clogging of the printer nozzle. The researchers could generate intricate patterns

© Springer International Publishing Switzerland 2017

R. Jelinek, *Carbon Quantum Dots*, Carbon Nanostructures,
DOI 10.1007/978-3-319-43911-2_7

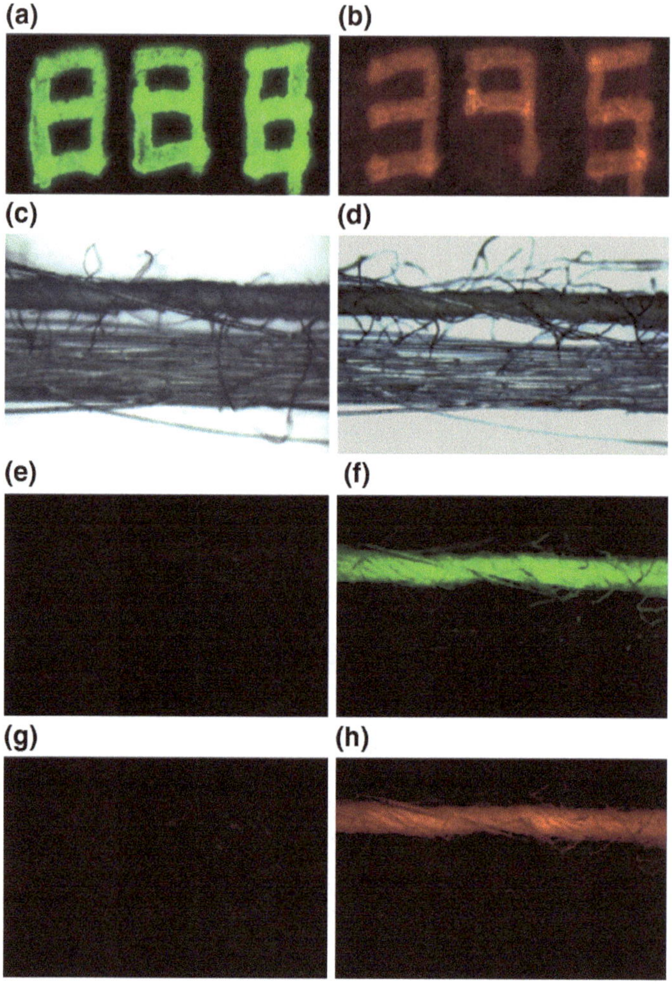

Fig. 7.1 Carbon dot "invisible ink" printed on organic materials. **a–b** Printing using aqueous solution containing commercially-available green fluorescent dye and Carbon-Dots on conventional paper. Emission filter 450–480 nm (UV excitation) (**a**) shows the fluorescence of the green dye, while emission filter 510–550 nm (excitation at 500 nm) reveals the fluorescence of the Carbon-Dot "ink" (**b**). **c**, **e**, **g** Optical and fluorescence images of a cotton fiber (*top*, **c**) and nylon fiber (*bottom*, **c**) unlabeled with Carbon-Dots. **d**, **f**, **h** Optical and fluorescence images of a cotton fiber (*top*, **d**) and nylon fiber (*bottom*, **d**) *labeled* with Carbon-Dots. Fluorescence of *only* the cotton fiber is apparent, reflecting attachment of the Carbon-Dots only to the organic fiber (and not to the nylon fiber). Reprinted from Qu et al. [1], with permission from John Wiley & Sons (2012)

using the Carbon-Dot/glycol ink using a conventional inkjet printer. Importantly, the printed patterns exhibited luminescence upon illumination by UV light, indicating that the Carbon-Dots retained their optical properties following drying and deposition upon the printed surfaces. The researchers further showed that different luminescence colors could be obtained upon co-addition of fluorophores (fluorescent dyes or semiconductor nanoparticles) to the Carbon-Dot-based ink. The color changes in such mixtures were ascribed to energy transfer between the Carbon-Dots and the co-added fluorophores (akin to FRET processes).

The broad excitation/emission range of Carbon-Dots and high quantum yield obtained in many instances have attracted interest for possible use of these nanoparticles in *solar energy* applications. Specifically, Carbon-Dots have been proposed as *sensitizers* in dye-sensitized solar cells (DSSCs). The operation principles of DSSCs are outlined in Fig. 7.3. TiO_2 *nanoparticles* constitute a core component in many DSSCs. Since bare TiO_2, which is a wide bandgap semiconductor, absorbs energy in the ultraviolet region of the electromagnetic spectrum, the applicability of the pure material is limited. To overcome this limitation, *doping* of the TiO_2 with other molecular species, referred to as *sensitizers*, is generally carried out. The sensitizers have *lower energy* bandgaps and thus absorb light in the visible (or infrared) regions, transferring the excited electrons into the conduction band of TiO_2 (Fig. 7.3a).

The basic DSSC configuration is shown in Fig. 7.3b. Light is absorbed by sensitizers attached to the TiO_2 NPs. The photoexcited electrons are transferred to the conducting band of the TiO_2 (according to the mechanism outlined in Fig. 7.3a) and transported to one of the electrodes. In parallel, reduced/oxidized (redox) pairs

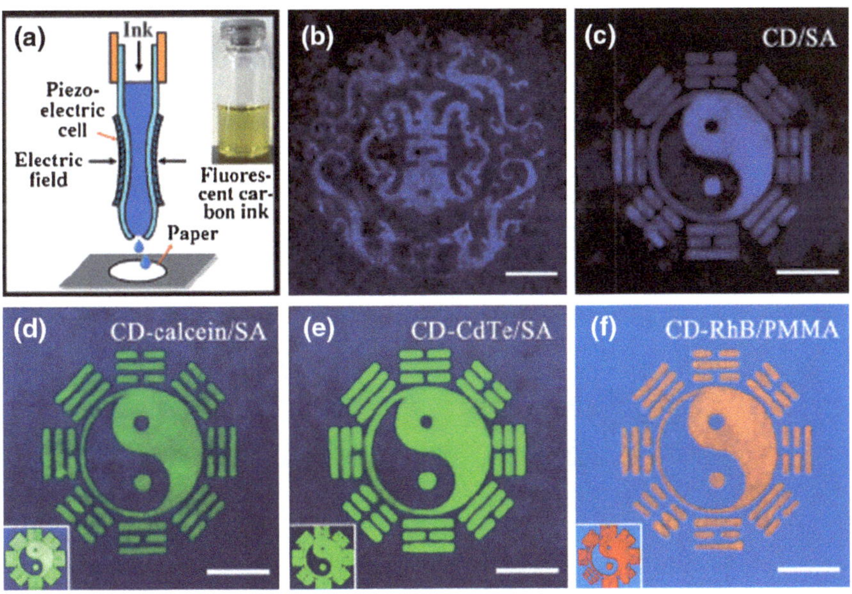

Fig. 7.2 Inkjet printing of carbon dot/glycol ink. **a** Scheme of the inkjet printer nozzle, the inset on the right shows the Carbon-Dot ink solution. **b** Printed pattern illuminated with UV excitation. **c–f** A "Tai-chi" pattern printed using mixtures of Carbon-Dots and different fluorophores. Sodium alginate (SA) is used as a stabilizer for the Carbon-Dots. The *insets* at the *bottom left* in **d–f** correspond to the fluorescent patterns of the fluorophore/SA ink (without Carbon-Dots). Scale bars correspond to 1 cm. Reprinted from Wang et al. [2], with permission from John Wiley & Sons (2012)

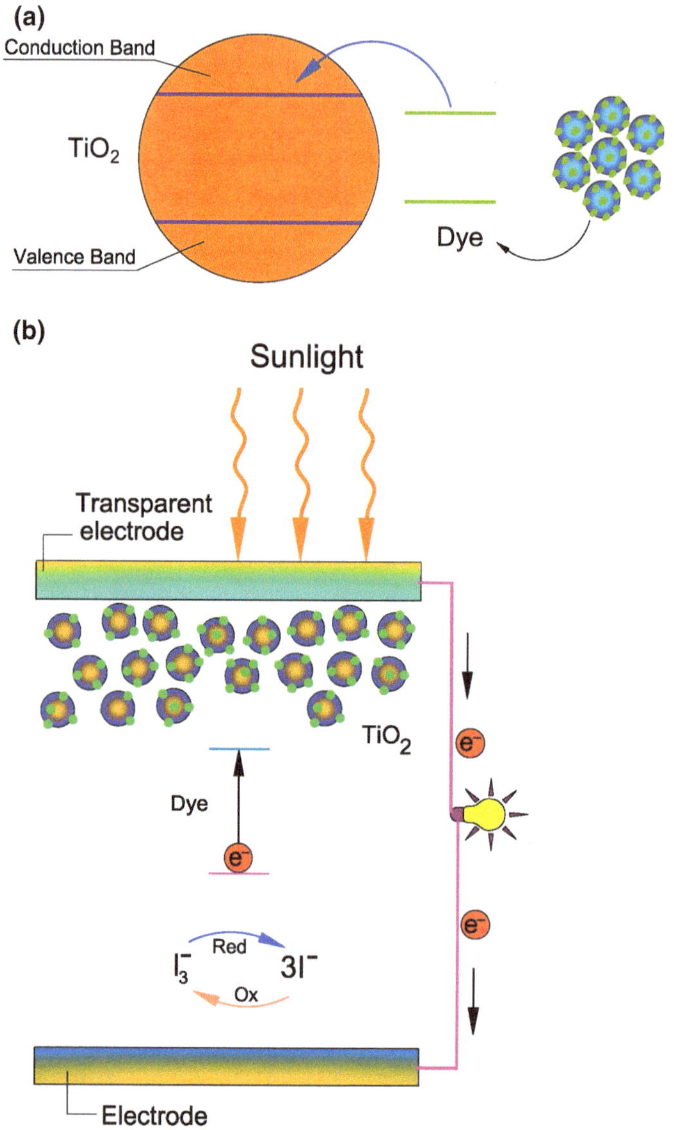

Fig. 7.3 Operation principle of dye-sensitized solar cell (DSSC). **a** Energy levels of TiO_2 particle and the dye; the energy bandgap of the dye is lower and thus enables absorption in the visible region of the electromagnetic spectrum. **b** Scheme depicting the complete electron cycle in a DSSC

(usually I^-/I_3^-, also denoted as the *electrolytes*) embedded in the cell are responsible for providing the electrons to reduce the sensitizer back to its initial state, in parallel receiving the electrons from the back electrode, thereby closing the electrical circuit. L. Li and colleagues at Indiana University have demonstrated that graphene quantum dots (QGDs) could serve as sensitizers in DSSCs [3]. The researchers synthesized relatively bulky (~15 nm) GQDs, stabilized by trialkylphenyl moieties (Fig. 7.4a), which were readily solubilized in common organic solvents.

The GQDs outlined in Fig. 7.4 were amenable to DSSC applications. Specifically, the particles displayed an absorption peak at around 600 nm (Fig. 7.4b)—well within the visible spectral region, thus could efficiently absorb sunlight energy. The current/voltage profile of a DSSC in which the GQDs served as sensitizers confirmed functionality of the cell. Figure 7.4c shows significant current generated upon illumination. The DSSC design illustrated in Fig. 7.4 is notable, because realizing the potential of *graphene* in solar cells applications has been so far limited by the low solubility and aggregation of graphene sheets. As such, GQDs might open promising avenues to utilize graphene in solar cells.

Carbon-Dots have been utilized as core components in other types of solar cell designs. Figure 7.5 depicts a "bulk heterojunction" (BHJ) organic photovoltaic cell in which GQDs were embedded in the BHJ medium, contributing to both light energy absorbance and charge mobility and thus enhancing the cell performance [4]. Interestingly, in that study B.H. Hong and colleagues at Seoul National University, Korea, demonstrated that the degree of *oxidation/reduction* of the GQDs intimately affected the solar cell's properties. In particular, the researchers

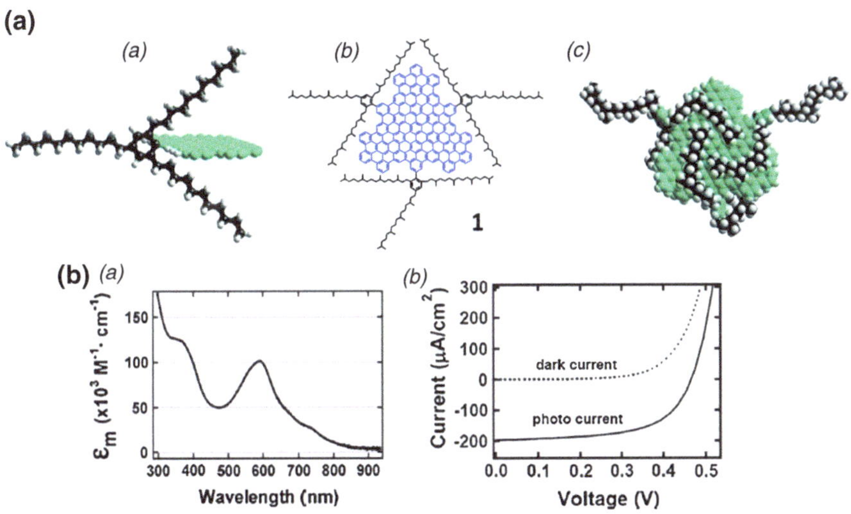

Fig. 7.4 Graphene quantum dots as sensitizers in dye-sensitized solar cells. **a** Schematic structure of the solubilized GQDs. The GQDs are encapsulated by trialkylphenyl residues which promote solubilization of the particles in organic solvents. *a* Drawing indicating bonding between the phenyl moiety and graphene; *b* stabilization of the GQD core by the alkyl chains; *c* theoretical calculation depicting the encapsulation of a GQD by three trialkylphenyl residues. **b** Photophysical properties of the solubilized GQDs. *a* UV-vis absorbance spectrum showing a prominent absorbance peak at 600 nm (visible light); *b* photocurrent generated by a DSSC comprising the GQDs. Reprinted with permission from Yan et al. [3]. Copyright (2010) American Chemical Society

reported that *intermediate* reduction levels of the embedded GQDs produced the most optimal cell performance. This result was ascribed to the increased abundance of the *carbene* residues, formed through the reduction process, which exhibit a favorable excitation in the UV range. (However, a presumed secondary energy transition associated with *CO moities* was adversely affected by excessive reduction, accounting for the observation that intermediate reduction was the most favorable condition). Overall, the apparent relationship between reduction time and the energy profile of the photovoltaic cell confirmed the significant role of the GQDs in affecting photovoltaic cell performance.

The light absorption properties of Carbon-Dots have been exploited as means to enhance ***photocatalysis***. Photocatalytic reactions have been studied in most cases through the use of metal oxide semiconductors such as titanium oxide and silicon oxide, particularly in *nanoparticle* configurations [providing a high surface area]. The main impediment, however, for using many metal oxide configurations as photocatalysts is their wide energy bandgap, which translates into inefficient absorbance of sunlight. Accordingly, common methods to enhance photocatalytic activity of metal oxide have been to couple them to dyes or nanoparticles which do absorb visible light and subsequently convert it to higher energy emission (e.g., shorter wavelengths).

Carbon-Dots have been proposed as promising candidates for photocatalytic enhancement through coupling to semiconductor substances. Two primary mechanisms of Carbon-Dot-enhanced photocatalysis have been outlined (Fig. 7.6). One prominent pathway consisted of using the Carbon-Dots as light absorbers in the visible region, consequently transferring the photoexcited electrons to the semiconductor particles for catalytic action (Fig. 7.6a). An alternative route is employing the Carbon-Dots as "electron sinks"—accepting photoexcited electrons

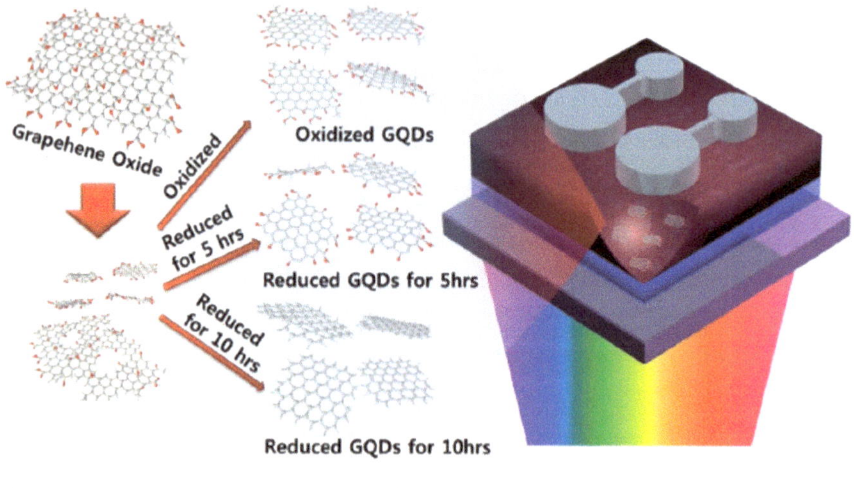

Fig. 7.5 Photovoltaic cell containing reduced graphene quantum dots. Cell performance found to correlate with the extent of GQD reduction; *intermediate reduction* ($\sim$5 h) found to produce the most optimal conditions. Reprinted with permission from Kim et al. [4]. Copyright (2013) American Chemical Society

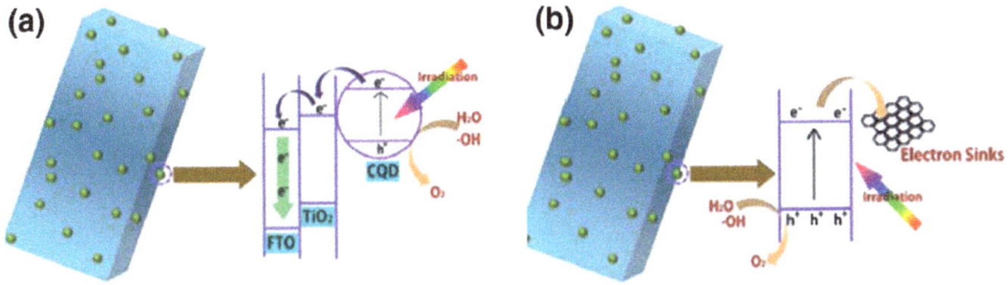

Fig. 7.6 Mechanisms of photocatalysis enhanced by carbon-dots coupled to a semiconductor material. **a** Carbon-Dots (*small green spheres*) act as light absorbers, transferring the photoexcited electrons to the semiconductor; **b** Carbon-Dots serve as "electron sinks" thereby extending the lifetimes of the electron-hole pairs. Reproduced from Xie et al. [18], with permission of the Royal Society of Chemistry

from the semiconductor material and thus minimizing electron-hole recombination, thereby extending catalytic activity (Fig. 7.6b). In both scenarios, the use of Carbon-Dots is favored since they exhibit broad and tunable excitation ranges (which in most cases include the *visible* spectral region). Furthermore, in many instances Carbon-Dots display efficient luminescence emission (an important parameter which affects overall photocatalytic performance). In addition, the Carbon-Dots' nanoscale size regime aids their incorporation within porous catalytic systems (considered beneficial since they exhibit high surface areas).

An experiment consistent with the energy transfer photocatalysis model (e.g., Fig. 7.6a) was reported by S.T. Lee and colleagues at Hong Kong SAR [5]. As depicted in Fig. 7.6, the researchers recorded significantly faster rate of photocatalytic degradation of an organic dye when Carbon-Dots were attached onto the surface of SiO_2 or TiO_2 nanoparticles. This result was ascribed to visible-light absorbance by the Carbon-Dots and subsequent electron transfer to the metal oxide nanoparticles (Fig. 7.7d); he transferred photoexcited electrons induced more pronounced degradation of the organic dyes.

In the example above, *upconversion* of visible light by Carbon-Dots appeared to be the mechanism responsible for the photocatalytic activity, transforming visible light adsorbed by the Carbon-Dots to *shorter* wavelength of the emitted light; it should be emphasized, however, that the occurrence of upconversion in Carbon-Dot systems has been recently disputed in several studies. There have been reports of effective photocatalysis accomplished through *downconversion*—a process in which light absorbed by Carbon-Dots is converted to *lower energy* emissions, which are then exploited for catalyzing target reactions. S.T. Lee and colleagues have shown that GQDs coupled to *rutile* TiO_2 particles—titanate crystalline phase which exhibit energy bandgap that was *lower* than the emission wavelength of the dots—featured an order-of-magnitude greater photocatalytic performance compared to a GQD/*anatase*-TiO_2, which has an energy gap that is *higher* than the GQDs emission (Fig. 7.8) [6]. This intriguing result essentially demonstrates that downconversion in GQD systems might enable, in some cases, efficient energy transfer available for photocatalysis.

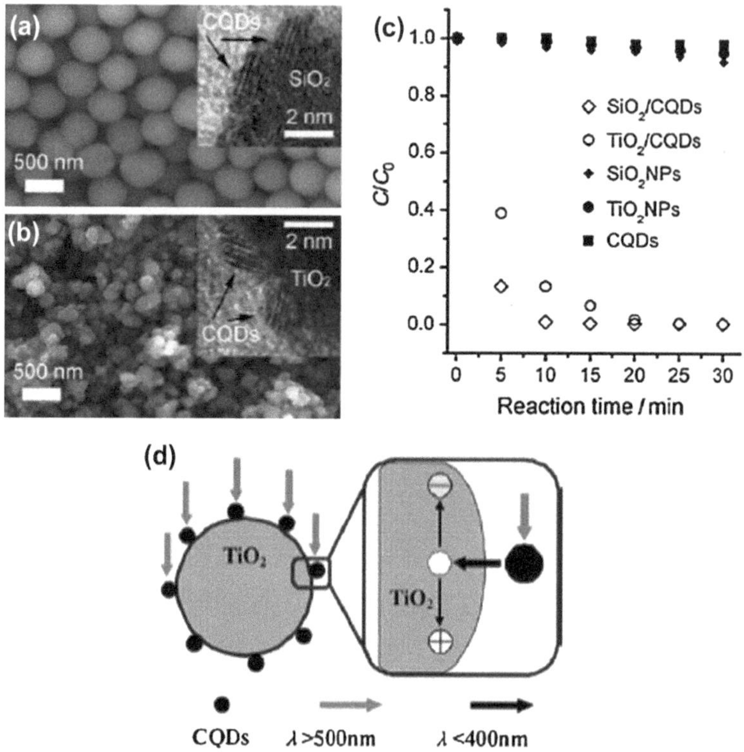

Fig. 7.7 Photocatalysis in a TiO_2/SiO–$_2$–—carbon dot system. **a–b** Scanning electron microscopy (SEM) images of SiO_2/Carbon-Dot and TiO_2/Carbon-Dot systems, respectively. The insets present the high resolution TEM images revealing the crystalline cores of the Carbon-Dots. **c** Enhanced photocatalytic activity in the particles containing Carbon-Dots. The *unfilled* datapoints reveal pronounced degradation of the organic dye attached to the particles. **d** Photocatalysis model: visible light absorbed by the Carbon-Dots and transferred to the metal oxide. Reprinted from Li et al. [5], with permission from John Wiley & Sons (2010)

Composite nanoparticles comprising Carbon-Dots coupled to metallic NPs, particularly *gold NPs*, have also exhibited catalytic properties. Z. Kang and colleagues at Soochow University, China, prepared Carbon-Dot/Au NPs and investigated their catalytic properties [7]. Interestingly, the researchers discovered that as-prepared Carbon-Dots (synthesized through electrochemical ablation of graphite) spontaneously reduced auric acid ($HAuCl_4$, the conventional starting material for Au NP synthesis), generating Au NPs. Moreover, as shown in Fig. 7.9, the Carbon-Dots were attached to the Au NPs; this feature prevented aggregation of the Au NPs in water. Particularly noteworthy was the observation of enhanced catalytic performance of the composite Carbon-Dot/Au NPs in decomposing organic compounds. This result was ascribed to pronounced adsorption of the target molecules, which was promoted by the Carbon-Dots that together with the Au NPs provided substantial surface area in the composite material.

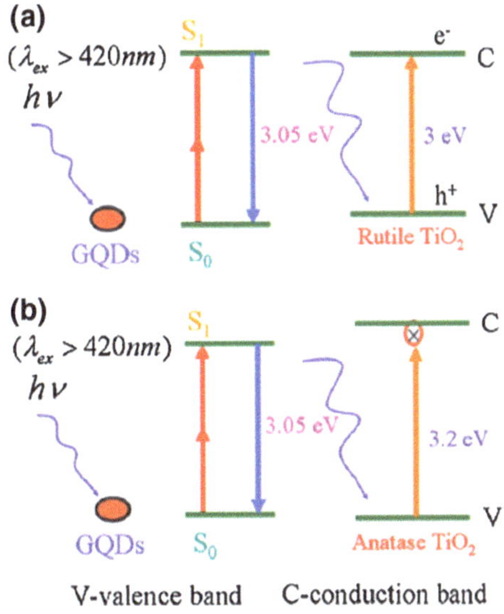

Fig. 7.8 Photocatalysis in composite materials comprising graphene quantum dots coupled to TiO_2 in different crystalline phases. Relative energy bandgaps and energy transfer from GQDs to rutile TiO_2 (**a**) or anatase TiO_2 (**b**). More pronounced catalysis was recorded in case of *rutile* TiO_2 for which the energy bandgap is smaller than the GQDS as compared to *anatase* TiO_2, which exhibits greater energy gap. Reprinted with permission from Zhuo et al. [6]. Copyright (2012) American Chemical Society

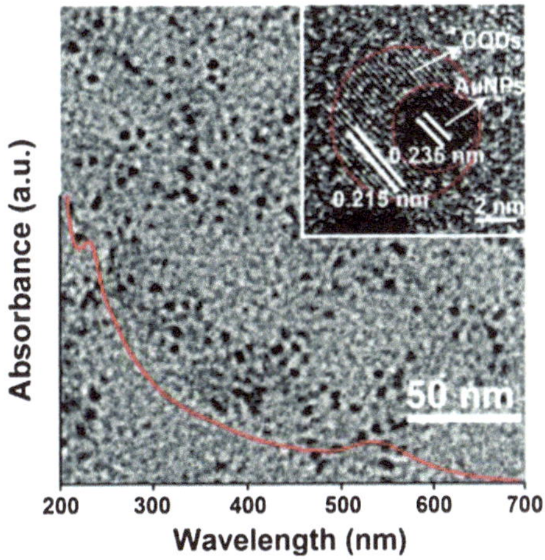

Fig. 7.9 Composite carbon dot/gold nanoparticles. A TEM image of the particles; the high resolution TEM (HRTEM) image in the inset confirms association between the crystalline Au NP and Carbon-Dot. The UV spectrum of the composite material, shown in red, reveals the plasmon resonance of gold around 550 nm. Reproduced from Liu et al. [7], with permission of the Royal Society of Chemistry

Another demonstration of enhanced catalytic properties of a Carbon-Dot/Au NP composite system is depicted in Fig. 7.10. This study, conducted by B.K. Gupta and colleagues at CSIR, India, explored the catalytic activity for reduction/decomposition of nitrophenol compounds by substances comprising anisotropic (rod-like) Au nanostructures coupled to carboxylate-displaying Carbon-Dots [8]. As shown in Fig. 7.10 (bottom), the conjugated particles were assembled through hydrophobic interactions between long hydrocarbon chains coating the Carbon-Dots and surfactants constituting the protective layer upon the Au nanorods. The more pronounced catalytic action of the Carbon-Dot/Au NP constructs was ascribed to both electron transfer from the Carbon-Dots to active sites upon the Au surface, as well as promoting

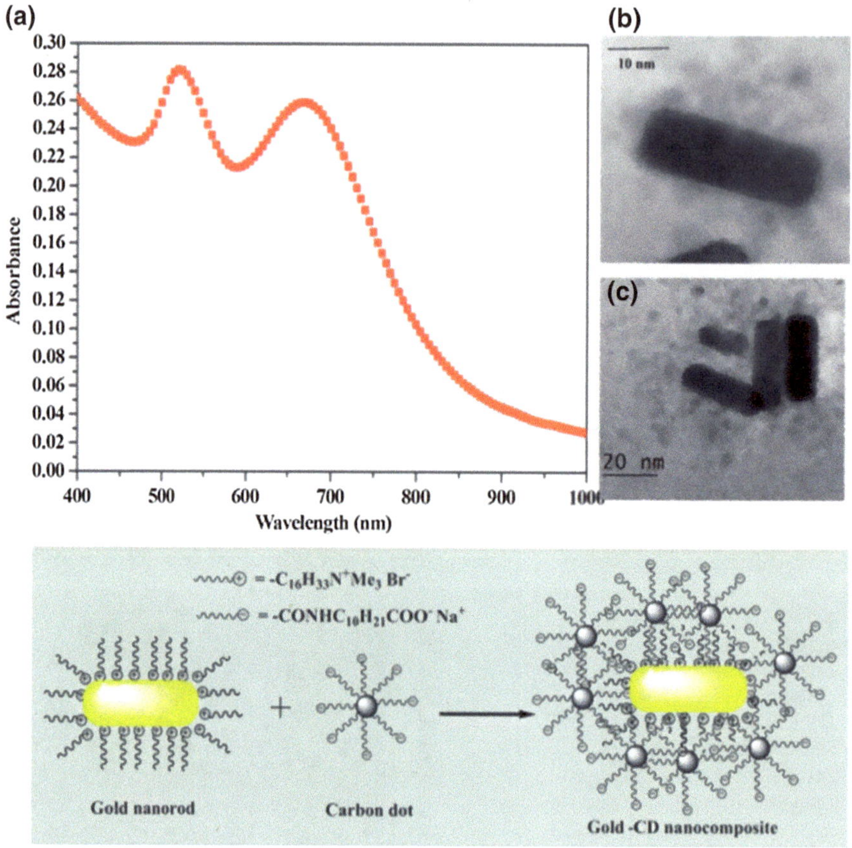

Fig. 7.10 Gold nanorod/carbon dot conjugates. **a** The composite nanoparticles retain the transverse and longitudinal plasmon resonance peaks associated with the Au nanorods (NRs); **b–c** the TEM images show the rod structures. The bottom scheme illustrates the proposed assembly mechanism, induced by the amphiphilic layers of both Au NPs and Carbon-Dots. Reproduced from Mondal et al. [8], 25863, with permission of the Royal Society of Chemistry

adsorption of the target nitrophenol molecules by hydrophobic domains present in the Carbon-Dots anchored to the gold.

Figure 7.11 illustrates a *Carbon-Dot/*Fe_2O_3 composite system exhibiting enhanced photocatalytic properties. The material was synthesized by Z. Kang and colleagues at Soochou University, China, via hydrothermal reaction of Carbon-Dots and the precursors for the iron oxide particles, yielding cubic Fe_2O_3 particles with Carbon-Dots attached onto their external facets [9]. The Carbon-Dot/Fe_2O_3 nanoparticles exhibited enhanced photocatalytic activity towards degradation of volatile hydrocarbon gases compared to bare iron oxide colloids (that were not coupled to Carbon-Dots). The researchers hypothesized that the enhanced photocatalysis was due to "storage" by the Carbon-Dots of the photoexcited electrons originating from the Fe_2O_3 particles (Fig. 7.11b). "Shuttling" of the electrons to the Carbon-Dots increased the lifetime of the

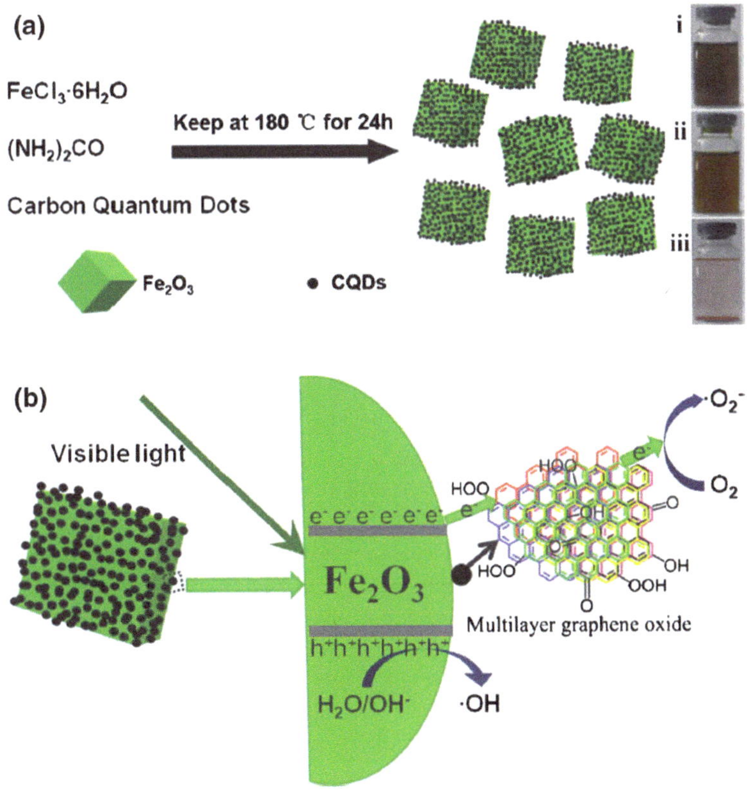

Fig. 7.11 Composite Fe_2O_3/carbon-dots and their application in photocatalysis. **a** Synthesis of the material through hydrothermal treatment of Carbon-Dots together with the iron oxide precursors. The photographs on the right confirm occurrence of the reaction: *i* and *ii* correspond to the Carbon-Dot solutions prior to the hydrothermal treatment while *iii* shows disappearance of the deep color of the soluble Carbon-Dots due to formation of the Fe_2O_3/Carbon-Dots. **b** Proposed mechanism of Carbon-Dot-enhanced photocatalysis. Electrons are transferred to the Carbon-Dots, the radicals generated by the electrons and holes oxidize and degrade the adsorbed organic molecules. Reproduced from Zhang et al. [9], with permission of the Royal Society of Chemistry

electron-hole pairs, thus contributing to more effective reactivity of the iron oxide in degradation of the adsorbed organic molecules.

While photocatalysis has been demonstrated in composite systems comprising Carbon-Dots and varied semiconducting and metallic substances such as ZnO, TiO_2, SiO_2, Fe_2O_3 and others, several reports have shown that Carbon-Dots by themselves exhibit photocatalytic activities. Indeed, the photoexcited electrons (and corresponding holes) in Carbon-Dots can be exploited for inducing diverse catalytic reactions upon the nanoparticles' surface, similar to the common scenarios of photocatalysis in *semiconductor* NP systems. Figure 7.12 illustrates representative generic processes in which electrons/holes generated by sunlight illumination upon the Carbon-Dot's surface participate in photocatalytic reactions.

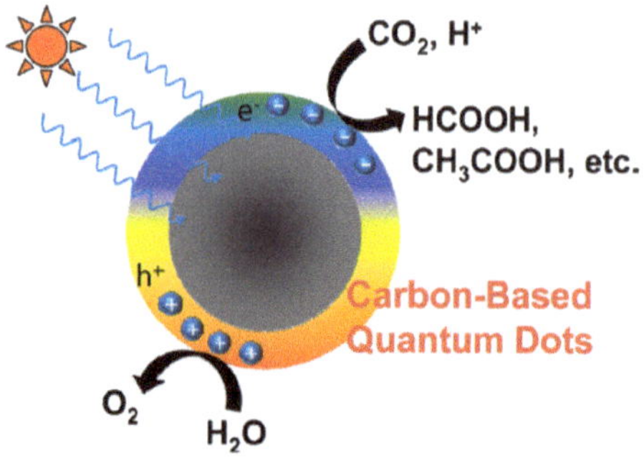

Fig. 7.12 Photocatalysis achieved by carbon-dots. The electrons and holes generated by light participate in catalytic reactions such as water oxidation and carbon dioxide degradation. Reprinted with permission from Fernando et al. [19]. Copyright (2015) American Chemical Society

The color properties of Carbon-Dots have naturally attracted interest for potential *optical* and *photonic* applications. Such uses, however, have faced significant challenges because of the need to employ Carbon-Dots in the *solid state*, rather than in solution. Specifically, when Carbon-Dots are immobilized or deposited upon substrates following solvent evaporation, they generally tend to aggregate and/or lose their luminescence properties. In particular, *fluorescence quenching* processes become a major problem in such cases.

Several reports presented innovative strategies aimed to overcome the above-mentioned aggregation/quenching hurdles. Y. Lin and colleagues at the South China Agricultural University have produced "quenching-resistant" Carbon-Dot powders in which the dots retained their bright intensities and color tunability [10] (Fig. 7.13). The researchers accomplished this by synthesizing nitrogen-doped Carbon-Dots from ethylenediamine and poly(vinyl-alcohol) (PVA). As a result, the Carbon-Dots were coated by a PVA layer, which upon drying of the particles prevented luminescence quenching through maintaining a large enough separation between the immobilized nanoparticles, thus blocking non-radiative energy transfer processes (Fig. 7.13a). Interestingly, using *different* nitrogen sources in the preparation procedure, Carbon-Dot powders exhibiting distinct colors were prepared. The powders additionally had different colors under conventional daylight illumination and upon irradiation by UV light (Fig. 7.13b).

Another strategy for fabrication of Carbon-Dot-based luminescent devices relied upon a one-pot synthetic process in which flexible polymer films containing immobilized Carbon-Dots were synthesized (Fig. 7.14) [20]. The single-step preparation scheme consisted of mixing the precursor substances for *both* polymer (polydimethylsiloxane, PDMS) *and* Carbon-Dots. Gradual increase of the temperature yielded composite films comprising Carbon-Dots embedded within the polymer matrix. Importantly, the physical separation between the polymer-embedded Carbon-Dots ascertained that no aggregation-induced quenching occurred. Furthermore, distinct luminescence colors could be generated simply by preparing the Carbon-Dot/polymer films using different carbon precursors (forming Carbon-Dots exhibiting different fluorescence emission profiles), or using *mixtures* of such Carbon-Dot precursors. Indeed, embedding Carbon-Dot *mixtures* within the transparent PDMS film also made possible generation of *white light* by irradiation with a blue-emitting light-emitting diode (LED); the mixture of Carbon-Dot-emitted wavelengths provided the chromatic components of white light.

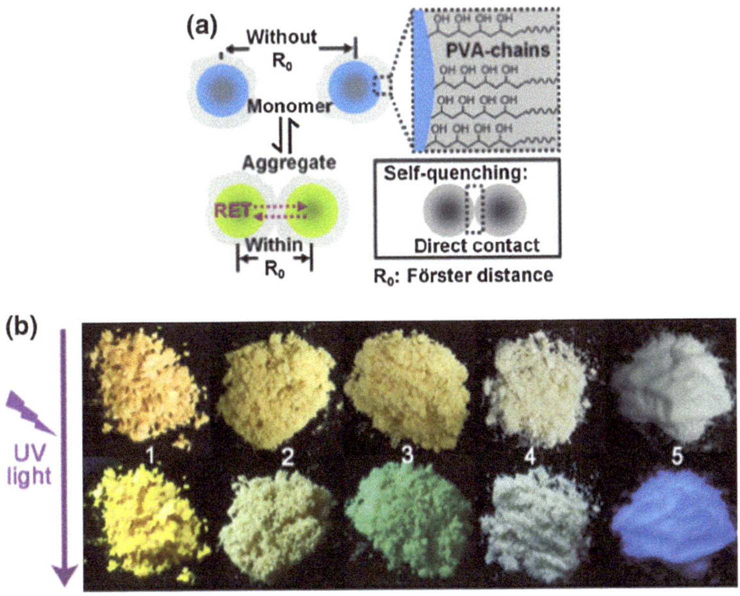

Fig. 7.13 Carbon dot powders having different colors. The Carbon-Dots were prepared from nitrogen-containing compounds and poly(vinyl-alcohol) (PVA). **a** Illustration showing that the PVA layer inhibits self-quenching by maintaining a large distance between the Carbon-Dots; **b** Powders comprising Carbon-Dots produced from PVA and different nitrogen-containing compounds (compounds 1-5). Note that different colors are apparent both with (bottom row) and without UV illumination (top row). Reprinted from Chen et al. [10], with permission from John Wiley & Sons (2016)

Carbon-Dot-based white light emitters using other composite materials have been demonstrated. Figure 7.15 illustrates a transparent film comprising Carbon-Dots and lanthanide complexes acting as the phosphors, all embedded within a transparent polymer matrix. This light emitter, designed by J. Feng and colleague at Fudan University, China, relied on Carbon-Dots serving as the *blue* phosphors, $Eu(DBM)_3$ (DBM = dibenzoylmethide) constituting the *red* phosphor, and $Tb(DBM)_3$ as the *green* phosphor [11]. Importantly, all three light components were chemically functionalized to facilitate their covalent integration into the polymeric network of polymethoxymethacrylate (PMMA), a transparent polymer. Immobilization of the chromatic species within the polymer matrix maintained their stable light emission, preventing energy transfer processes among them and consequent quenching; indeed, the composite Carbon-Dot/lanthanide-complex/polymer matrix produced intense white light upon UV irradiation (Fig. 7.15b).

Other innovative approaches have been introduced to utilize Carbon-Dots in light-emitting devices. H. Lin and colleagues at the Institute of Materials Technology & Engineering (NIMTE), Ningbo, China succeeded to both produce Carbon-Dots with tunable colors, as well as immobilize the dots within transparent films, thereby generating a variety of visible colors [12]. Specifically, as shown in Fig. 7.16, the researchers discovered that construction of Carbon-Dots from different

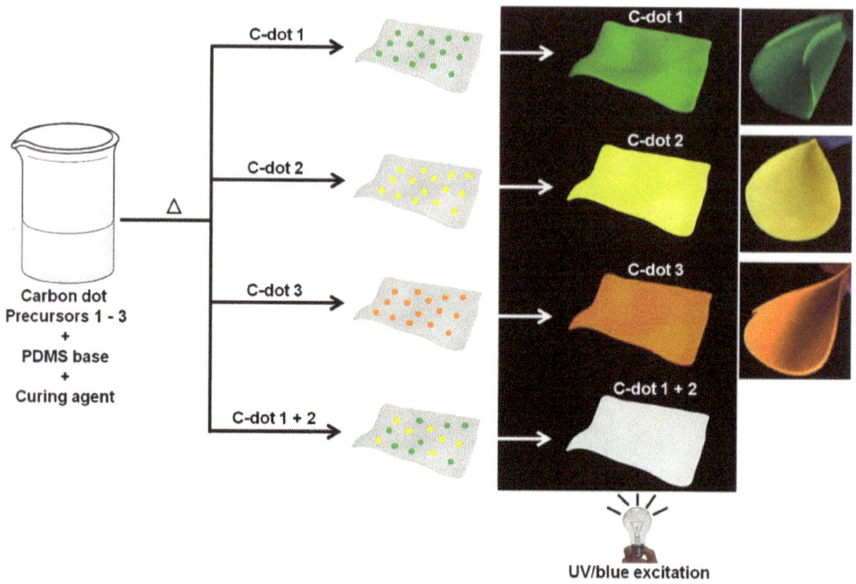

Fig. 7.14 Carbon-dots encapsulated in flexible transparent polymer light emitters. Different light colors are emitted (upon blue/UV irradiation), depending upon the type of Carbon-Dots synthesized within the polymer film. The column on the *right* shows photographs of the actual films upon illumination with a UV light-emitting diode (LED). Adapted from Bhunia et al. [20]

isomers of phenylenediamines through a simple hydrothermal method yielded Carbon-Dots having different colors upon UV irradiation, depending upon the isomer used. In particular, the three "basic" colors—red, green, blue—could be produced. Subsequent encapsulation of the Carbon-Dots within a transparent polymeric film matrix yielded brightly colored films upon illumination with UV light. Notably, as apparent in Fig. 7.16, a wide variety of colors (including white light) were generated by embedding *mixtures* of different Carbon-Dots within the polymer films.

An important prerequisite for Carbon-Dot light modulators and emitters, apparent in the examples discussed above, is *immobilization* of Carbon-Dots within

Fig. 7.15 White light emitter comprising a composite carbon dot/lanthanide complexes ▶ embedded in a transparent polymer film. **a** Synthesis scheme; functionalization of the Carbon-Dots is carried out to facilitate encapsulation within the polymer matrix and prevent aggregation. **b** Optical properties: (*a*) Chromaticity coordinates obtained in different film compositions, shown on a conventional CIE 1931 chromaticity chart (reflecting the color and quality of light emitted); the film indicated by the arrow emit white light. *b* Photograph of the white light-emitting film, upon irradiation with 400 nm laser. *c* Luminescence emission of the three individual molecular phosphors upon 400 nm excitation, and the emission spectrum of the white light-emitting film (black spectrum). Emission peaks of the composite film in three main spectral components (*red–green–blue*) underscore the white light emission. *d* Photograph of the transparent composite film in regular illumination. Reprinted with permission from Chen and Feng [11]. Copyright (2015) American Chemical Society

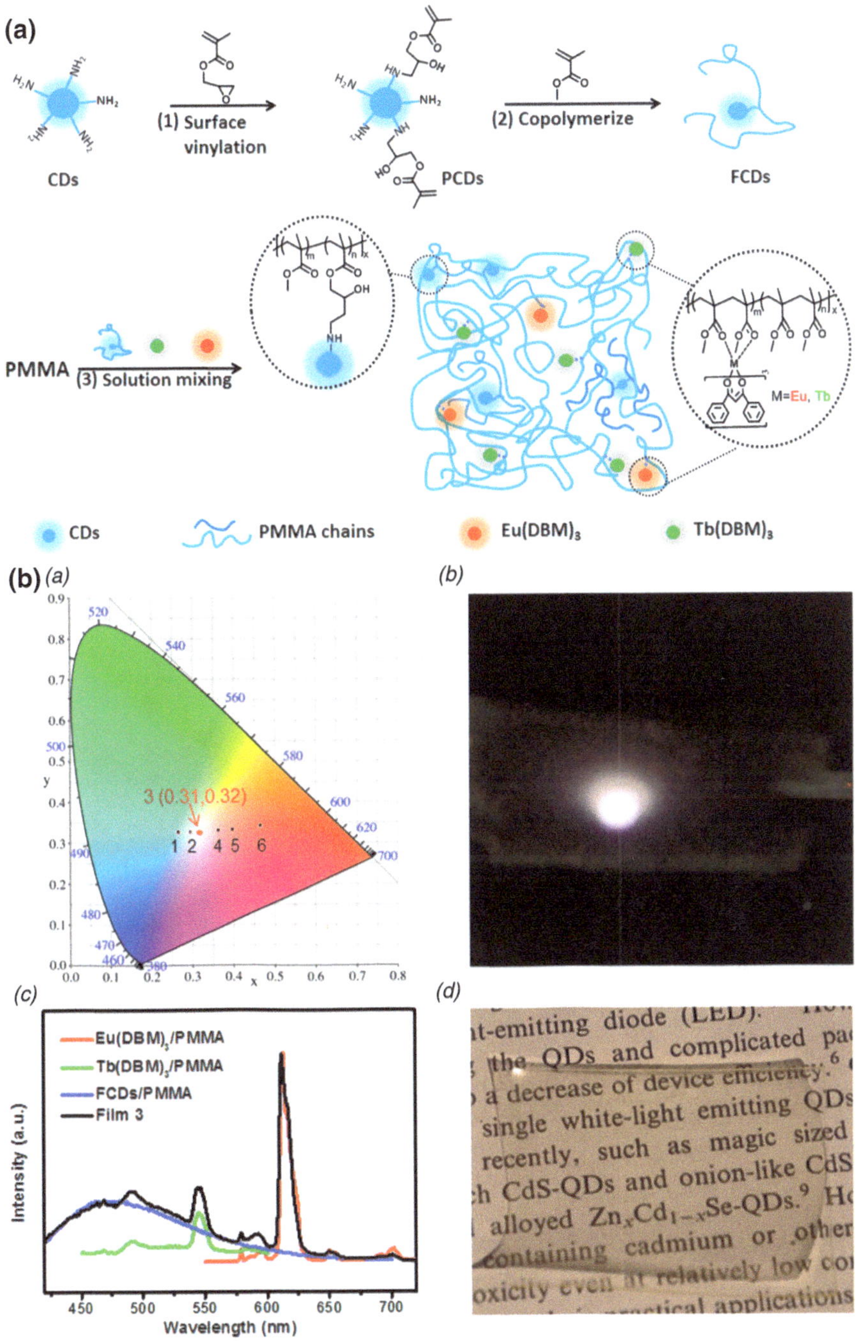
(a)
CDs
(1) Surface vinylation
PCDs
(2) Copolymerize
FCDs
PMMA
(3) Solution mixing
M=Eu, Tb
CDs
PMMA chains
Eu(DBM)3
Tb(DBM)3
(b)
(a)
(b)
3 (0.31,0.32)
1 2 4 5 6
x
y
(c)
(d)
Eu(DBM)3/PMMA
Tb(DBM)3/PMMA
FCDs/PMMA
Film 3
Intensity (a.u.)
Wavelength (nm)

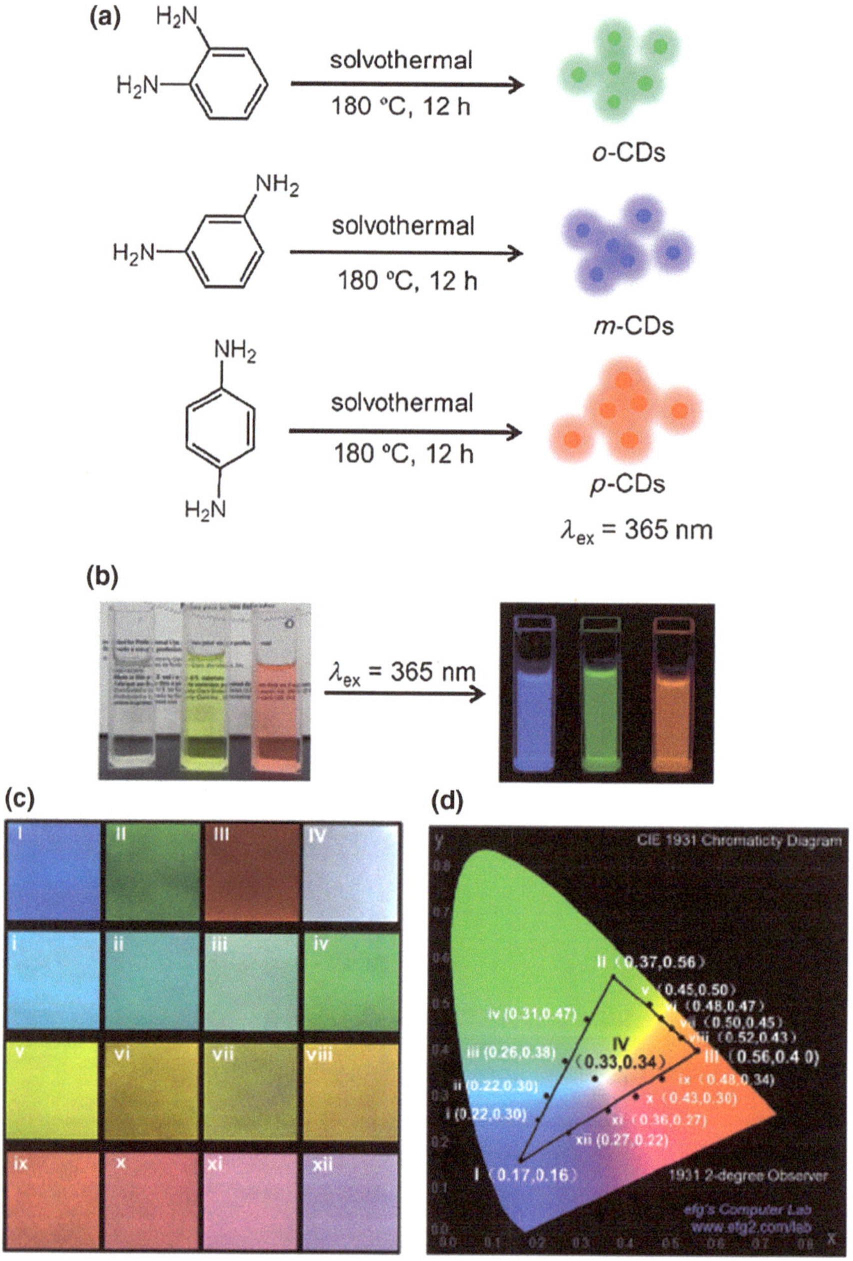
(a)
H_2N
H_2N
solvothermal
180 °C, 12 h
o-CDs
NH_2
H_2N
solvothermal
180 °C, 12 h
m-CDs
NH_2
solvothermal
180 °C, 12 h
H_2N
p-CDs
λ_{ex} = 365 nm
(b)
λ_{ex} = 365 nm
(c)
I
II
III
IV
i
ii
iii
iv
v
vi
vii
viii
ix
x
xi
xii
(d)
CIE 1931 Chromaticity Diagram
II (0.37,0.56)
v (0.45,0.50)
vi (0.48,0.47)
vii (0.50,0.45)
viii (0.52,0.43)
iv (0.31,0.47)
iii (0.26,0.38)
IV (0.33,0.34)
III (0.56,0.4 0)
ix (0.48,0.34)
x (0.43,0.30)
xi (0.36,0.27)
xii (0.27,0.22)
ii (0.22,0.30)
i (0.22,0.30)
I (0.17,0.16)
1931 2-degree Observer
efg's Computer Lab
www.efg2.com/lab

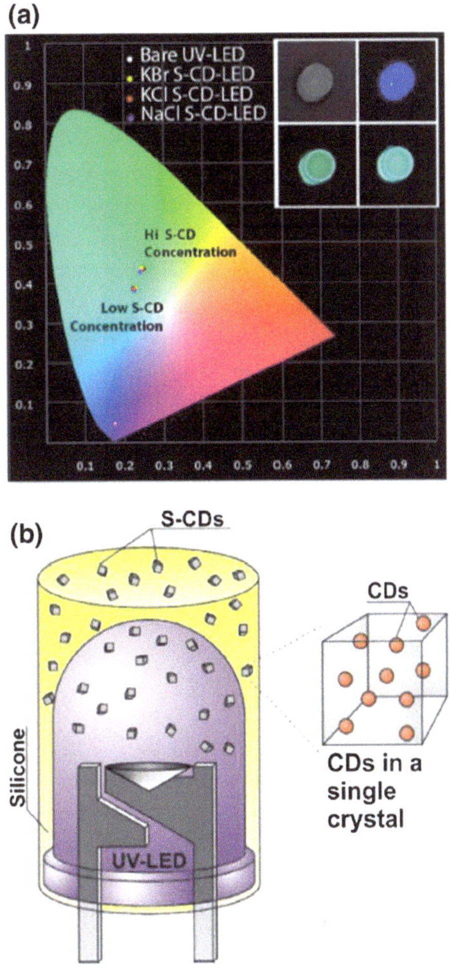

Fig. 7.17 Salt-embedded carbon-dots as conduits for light emitters. **a** Light coordinates in a CIE 1931 chromaticity map of emitters comprising Carbon-Dots embedded within different salt crystals; **b** schematic structure of the light-emitting device based upon the salt-embedded Carbon-Dots (denoted S-CDs). Reprinted from *J. Luminescence* **2014**, *154*, Kim et al. [13], Copyright (2014), with permission from Elsevier

solid matrixes, required both as means for practical use as well as in order to minimize aggregation-induced luminescence quenching. Another elegant example of an optical material featuring Carbon-Dots stabilized within a transparent matrix is depicted in Fig. 7.17. In that study, Q. Li and colleagues at Griffith University,

◀ **Fig. 7.16** Colored light-emitting films embedding Carbon-Dots prepared from phenylenediamine isomers in transparent polymer matrixes. **a** The three phenylenediamine isomers and corresponding Carbon-Dots exhibiting different colors (upon excitation at 365 nm); **b** the Carbon-Dot solutions shown in regular light and fluorescence recorded upon UV irradiation; **c** photographs of Carbon-Dot/polymer films comprising Carbon-Dots produced from different mixtures of the phenylenediamine isomers; **d** light emission coordinates of the films shown on a CIE 1931 chromaticity map. A film composition producing white light is indicated in bold. Reprinted from Jiang et al. [12], with permission from John Wiley & Sons (2015)

Australia, embedded Carbon-Dots within crystals of ionic *salts*, including KBr, NaCl, and others [13]. The crystalline matrix retained the luminescence properties of the Carbon-Dots, and even significantly aided their long-term physical and optical stability. The salt-encapsulated Carbon-Dots (denoted *S-Carbon-Dots*) could be further grounded; the powder could be then mixed with a rubbery polymer, yielding molded films that were placed upon UV light sources generating convertible LEDs the color of which depended upon the particular salt crystals employed for encapsulating the Carbon-Dots. The salt crystals in such applications might provide "shielding" for the embedded phosphors (i.e., the Carbon-Dots), as thermal/UV degradation is a well known problem in LED technologies.

Other studies combined chemical functionalization of Carbon-Dots with smart device designs to achieve tunable light emission. A. Rogach and colleagues at the City University of Hong Kong fabricated flexible films comprising silane-derivatized Carbon-Dots and an *ionic liquid* in which the emission color was, interestingly, dependent upon film *thickness* [14]. This light emitter design, illustrated in Fig. 7.18, combined several innovative strategies. First, the electrostatic interactions between the surface-displayed silane residues and ionic liquid molecules produced a flexible *ionogel* which could be mechanically manipulated (bended, stretched, etc.). Second, the researchers demonstrated that full color tuning was accomplished by modulating the ionogel film thickness; this phenomenon was ascribed to reabsorption/emission of light propagated through the film by the embedded Carbon-Dots.

In addition to serving as *converters* of light as illustrated in the applications discussed above, Carbon-Dots have been also employed as actual *light sources* in LED setups. A representative Carbon-Dot-based LED system is depicted in Fig. 7.19, in which A. Rogach and colleagues showed that light exhibiting different colors was generated by *electroluminescence*, i.e., placing the Carbon-Dots in an electric field [15]. As illustrated in Fig. 7.19, the assembled device containing an emission-producing layer of Carbon-Dots was placed between electron- and hole-transport layers; light was emitted by the Carbon-Dots upon current passage. Notably, the *color* emitted by the Carbon-Dots could be controlled by the voltage applied. Moreover, color and chromatic properties (i.e., degree of "warmth" of generated light) could also be tuned by modulating the device structure, particularly *thickness* of the Carbon-Dot layer, pointing to the possible use of this design in tunable-color LEDs.

An application of Carbon-Dots as a vehicle for evaluating the *critical micelle concentration* (*CMC*) of surfactant molecules has been recently reported (Fig. 7.20). In that study, H.F. Wu and colleagues at National Sun Yat-Sen University, Taiwan, demonstrated that fluorescence properties of water-soluble Carbon-Dots were sensitive to the presence of surfactants and their assembly states —*monomeric* or *micellar* structures [16]. In particular, the researchers discovered that the *Stokes shifts* associated with the Carbon-Dots' fluorescence (i.e., the wavelength difference between the excitation and emission peaks) exhibited significant differences upon the transition of the surfactants from monomers to micelles. This phenomenon has been ascribed to the interactions between the surfactants' headgroups and functional groups (particularly *charged* residues) upon the

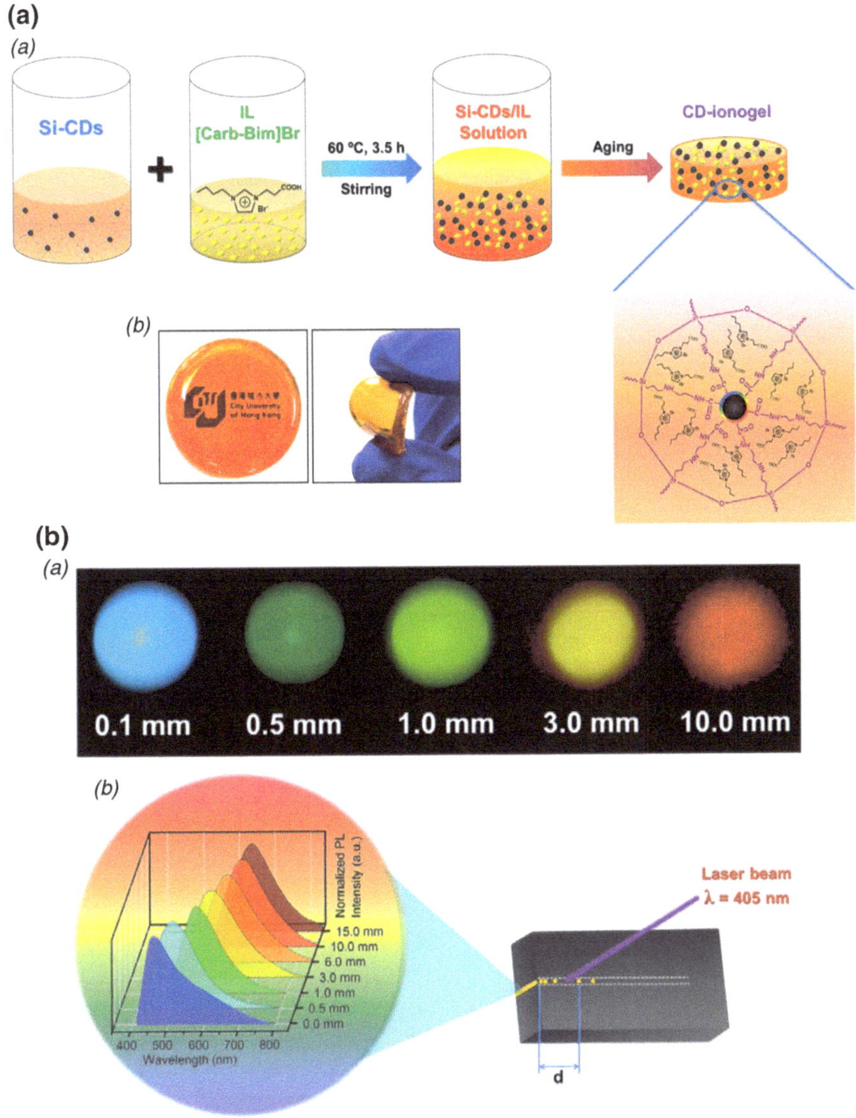

Fig. 7.18 Thickness-dependent colors in carbon dot/ionic liquid ionogel films. **a** Synthesis of the composite films from Carbon-Dot-silanes and ionic liquids. Photographs of the transparent and flexible ionogels are shown in *b*. **b** Thickness-dependent colors (*a*), and illustration of shift of the emission peak upon varying the thickness d (*b*). Reprinted with permission from Wang et al. [14]. Copyright (2014) American Chemical Society

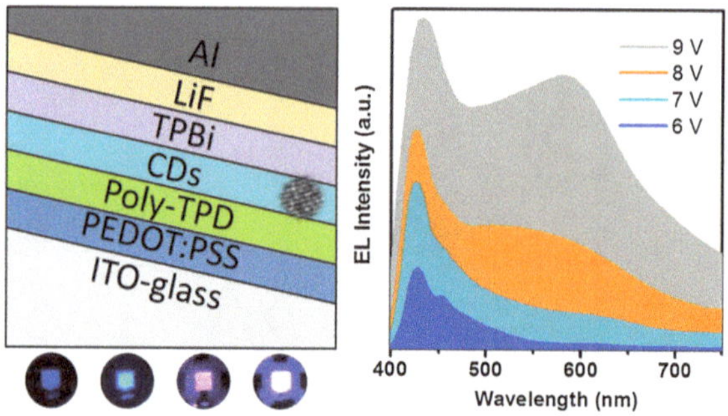

Fig. 7.19 Light-emitting diodes based on carbon-dots' electroluminescence. *Left* Schematic representation of the LED device, the light emissive Carbon-Dot layer is placed between the two electrodes. The photograph below shows light emitted upon application of different voltages. *Right* Electroluminescence spectra recorded in different voltage values. Reprinted with permission from Zhang et al. [15]. Copyright (2013) American Chemical Society

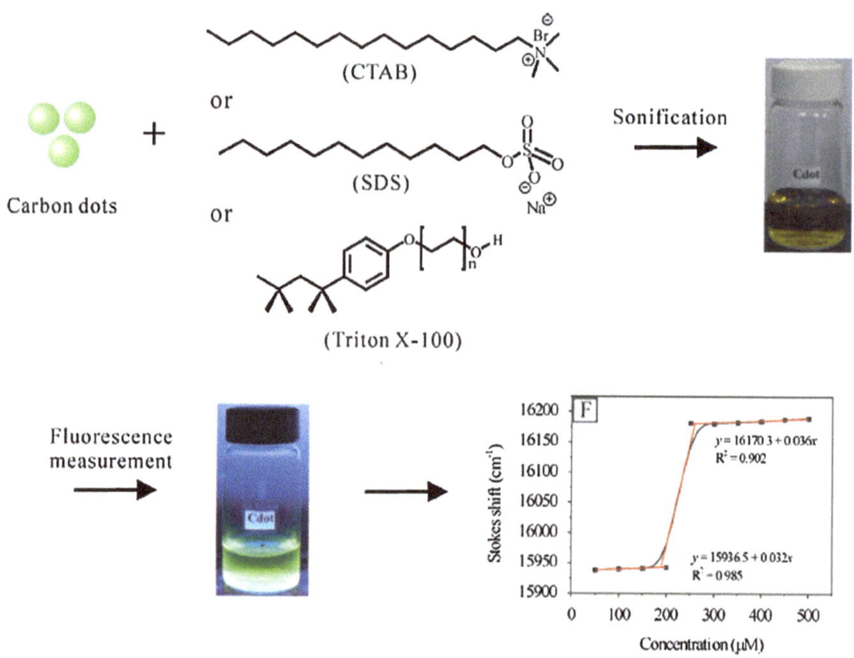

Fig. 7.20 Carbon-dots' fluorescence as a means for detection of critical micelle concentrations (CMCs). Scheme depicting the technique applied for detecting the CMCs of positive, negative, and neutral surfactants, respectively. The graph on the bottom right depicts the jump in the recorded fluorescence in the Carbon-Dot solution reflecting the CMC. Reprinted from *Talanta* **2015**, *132*, Bhaisare et al. [21], Copyright (2015), with permission from Elsevier

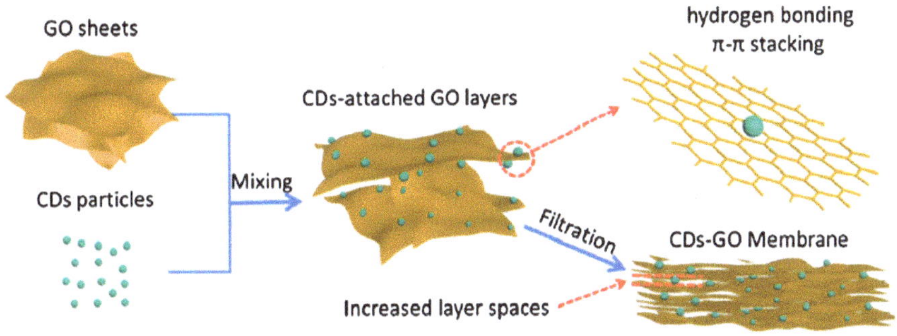

Fig. 7.21 Carbon dot/graphene oxide separation membranes. Scheme showing the membrane preparation process. The Carbon-Dots are inserted between the GO sheets, affecting membrane permeability. Reproduced from Wang et al. [17], with permission of the Royal Society of Chemistry

Carbon-Dots' surfaces; *micelles* presumably exhibited significantly different interactions with the Carbon-Dots compared to the soluble surfactant monomers. Indeed, the "jump" in the Stokes shift (Fig. 7.20) provided an excellent vehicle for identifying the CMC.

While the significant majority of Carbon-Dot applications have exploited the unique photophysical properties of these nanoparticles, some studies reported uses that do not relate to spectroscopic/optical aspects. Figure 7.21 illustrates construction of Carbon-Dot/graphene oxide (GO) *membranes*. In this composite system, Q. Li and colleagues at Griffith University, Australia, have shown that Carbon-Dots could be inserted between GO sheets, yielding a stacked porous structure exhibiting filtering capabilities [17]. The role of the Carbon-Dots in the assembly was important—maintaining nanometer-scale distance between adjacent sheets thus determining the filtration capacity of the membrane. Indeed, selecting Carbon-Dots at different sizes enabled tuning of membrane permeability.

References

1. S. Qu, X. Wang, Q. Lu, X. Liu, L. Wang, A biocompatible fluorescent ink based on water-soluble luminescent carbon nanodots. Angew. Chem. Int. Ed. **51**, 12215–12218 (2012)
2. J. Wang, C.-F. Wang, S. Chen, Amphiphilic egg-derived carbon dots: rapid plasma fabrication, pyrolysis process, and multicolor printing patterns. Angew. Chem. Int. Ed. **51**, 9297–9301 (2012)
3. X. Yan, X. Cui, B. Li, L.-S. Li, Nano Lett. **10**, 1869–1873 (2010)
4. J.K. Kim, M.J. Park, S.J. Kim, D.H. Wang, S.P. Cho, S. Bae, J.H. Park, B.H. Hong, ACS Nano **7**, 7207–7212 (2013)
5. H. Li, X. He, Z. Kang, H. Huang, Y. Liu, J. Liu, S. Lian, C.H.A. Tsang, X. Yang, S.-T. Lee, Water-soluble fluorescent carbon quantum dots and photocatalyst design. Angew. Chem. Int. Ed. **49**, 4430–4434 (2010)
6. S. Zhuo, M. Shao, S.-T. Lee, ACS Nano **6**, 1059–1064 (2012)

7. R. Liu, J. Liu, W. Kong, H. Huang, X. Han, X. Zhang, Y. Liu, Z. Kang, Dalton Trans. **43**, 10920–10929 (2014)
8. P. Mondal, K. Ghosal, S.K. Bhattacharyya, M. Das, A. Bera, D. Ganguly, P. Kumar, J. Dwivedi, R.K. Gupta, A.A. Marti, B.K. Gupta, S. Maiti, RSC Adv. **4**, 25863–25866 (2014)
9. H. Zhang, H. Ming, S. Lian, H. Huang, H. Li, L. Zhang, Y. Liu, Z. Kang, S.-T. Lee, Dalton Trans. **40**, 10822–10825 (2011)
10. Y. Chen, M. Zheng, Y. Xiao, H. Dong, H. Zhang, J. Zhuang, H. Hu, B. Lei, Y. Liu, A self-quenching-resistant carbon-dot powder withtunable solid-state fluorescence and construction of dual-fluorescence morphologies for white light-emission. Adv. Mater., **28**, 312–318 (2016)
11. B. Chen, J. Feng, J. Phys. Chem. C **119**, 7865–7872 (2015)
12. K. Jiang, S. Sun, L. Zhang, Y. Lu, A. Wu, C. Cai, H. Lin, Red, green, and blue luminescence by carbon dots: full-color emission tuning and multicolor cellular imaging. Angew. Chem. Int. Ed. **54**, 5360–5363 (2015)
13. T.H. Kim, F. Wang, P. McCormick, L. Wang, C. Brown, Q. Li, Salt-embedded carbon nanodots as a UV and thermal stable fluorophore for light-emitting diodes. J. Lumin. **154**, 1–7 (2014)
14. Y. Wang, S. Kalytchuk, Y. Zhang, H. Shi, S.V. Kershaw, A.L. Rogach, J. Phys. Chem. Lett. **5**, 1412–1420 (2014)
15. X. Zhang, Y. Zhang, Y. Wang, S. Kalytchuk, S.V. Kershaw, Y. Wang, P. Wang, T. Zhang, Y. Zhao, H. Zhang, T. Cui, Y. Wang, J. Zhao, W.W. Yu, A.L. Rogach, ACS Nano **7**, 11234–11241 (2013)
16. M.L. Lavkush, S. Pandey, M.S. Shahnawaz Khan, A. Talib, H.-F. Wu, Talanta **132**, 572–578 (2015)
17. W. Wang, E. Eftekhari, G. Zhu, X. Zhang, Z. Yan, Q. Li, Chem. Commun. **50**, 13089–13092 (2014)
18. Xie et al. J. Mater. Chem. A, **2**, 16365 (2014)
19. Fernando et al., ACS Appl. Mater. Interfaces **7**, 8363–8376 (2015)
20. Bhunia et al., Nanoscale **8**, 3400–3406 (2016)
21. Bhaisare et al. Fluorophotometric determination of critical micelle concentration (CMC) of ionic and non-ionic surfactants with carbon dots via Stokes shift, pp. 572–578 (2015)

Chapter 8
Carbon-Dot-Containing Composite Materials

The attractive chemical and photophysical properties of Carbon-Dots have provided impetus to research efforts aimed at creating mesoscale assemblies comprising Carbon-Dots as luminescent elements. Such materials also contain other chemical constituents having complementary roles, together producing multifunctional composite materials. This field of research has been aided by the chemical stability and resilience of Carbon-Dots to varied physical and chemical treatments.

Figure 8.1 schematically depicts a "carbon-dot composite nanoparticle" (CDCN) designed to function as an intracellular sensor for *glutathione* (*GSH*), an important antioxidant peptide. The CDCNs, constructed by W. Yang and colleagues at Fudan University, China, comprised of a *mesoporous silica nanoparticle* (*MSN*) framework, serving as a cell delivery vehicle of the Carbon-Dots, interspersed within a cross-linked disulfide-containing polymer (Fig. 8.1a) [1]. Initially, the Carbon-Dot aggregates did not fluoresce due to close proximity and self-quenching. However, as outlined in Fig. 8.1b, GSH molecules *cleaved* the sulfur–sulfur bonds within the Carbon-Dot embedding polymer matrix, thereby releasing the dots and consequently resurrecting their fluorescence emission. The researchers demonstrated that GSH sensing was accomplished intracellularly via this Carbon-Dot release mechanism.

While the Carbon-Dots in the CDCN assembly discussed above served as "passive" constituents—e.g., they did not directly respond to the external stimuli (the GSH molecules)—other composite materials employed Carbon-Dots as the actual targets for external triggering. An example of a *drug delivery* application utilizing Carbon-Dots as "gatekeepers" is shown in Fig. 8.2 [2]. The composite delivery system, designed by J. Ren and colleagues at the Changchun Institute of Applied Chemistry, comprised of MSNs acting as host matrix for the drug cargo. Importantly, the silica particles' surface was coated with positively charged residues. The Carbon-Dots were *negatively charged* as they contained abundant COO^- units, thus were electrostatically bound at the MSN surface serving as "caps" that prevented release of the encapsulated drug molecules. In mildly acidic conditions, however, the Carbon-Dots became protonated (as the COO^- residues formed neutral

© Springer International Publishing Switzerland 2017

R. Jelinek, *Carbon Quantum Dots*, Carbon Nanostructures,
DOI 10.1007/978-3-319-43911-2_8

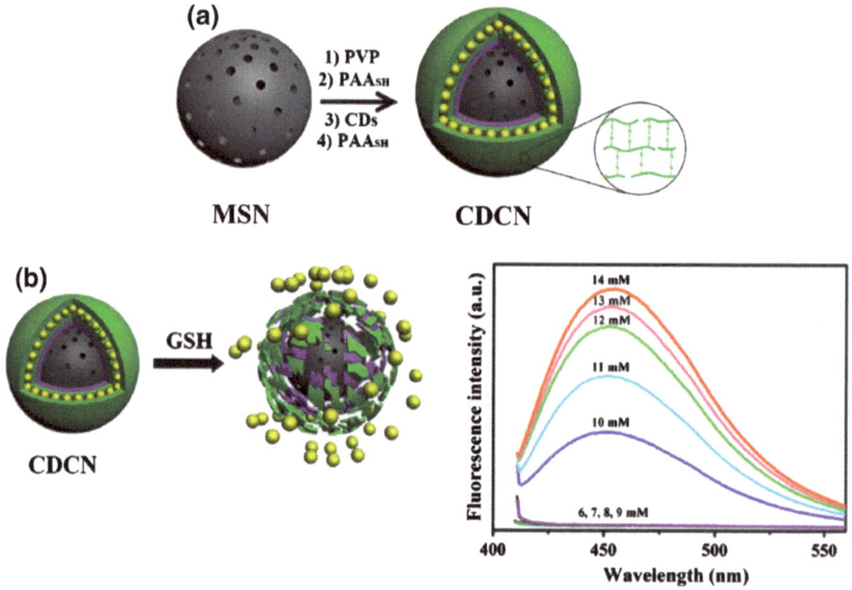

Fig. 8.1 Mesoporous silica nanoparticles (MSNs) embedding carbon-dots for detection of glutathione (GSH). **a** Synthesis and structure. The composite particles comprise of MSN encapsulating Carbon-Dots interspersed within a disulfide cross-linked polymer [thiolated poly (acrylic acid), PAA_{SH}]. **b** GSH breaks the disulfide bonds releasing the Carbon-Dots. Fluorescence spectra show the direct relationship between emission peak and GSH concentration. Reprinted from Liu et al. [1], with permission from John Wiley & Sons (2015)

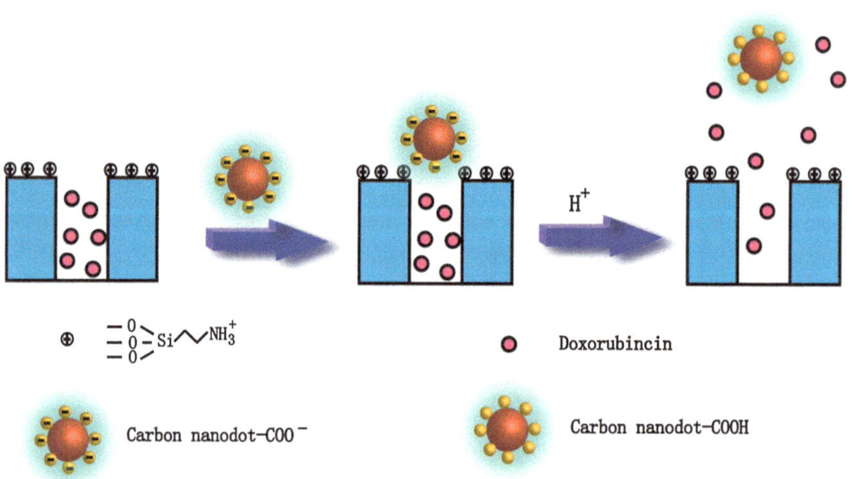

Fig. 8.2 Mesoporous silica nanoparticles (MSNs)/carbon-dot composites for controlled release of drug molecules. The Carbon-Dots are electrostatically attached onto the surface of MSNs. In acidic environments the Carbon-Dots are protonated and detached from the MSN matrix, inducing release of the doxorubicin drug molecules. Reprinted with permission from Zhou et al. [2]. Copyright (2013) American Chemical Society

COOH), consequently detaching from the MSNs and leading to release of the drug molecules. It should be also noted that the Carbon-Dots' *fluorescence* in the system allowed microscopic visualization of the delivery effectiveness.

Carbon-Dots have been recently employed as constituents in the expanding field of "theranostic materials"—multifunctional materials which exhibit both *therapeutic* and *diagnostic* properties. Theranostics rely on combining several components (primarily *nano*materials) into a single composite. Carbon-Dots have been mostly employed as *luminescent* species within theranostic materials, further taking advantage of their diverse functionalization routes and biocompatibility which contribute to their use in such applications. Figure 8.3 illustrates a representative theranostic assembly containing Carbon-Dots. The composite material was designed by S. Zhou and colleagues at the City University of New York, Staten Island to achieve multiple functions, including delivery of therapeutic substances, magnetic-field-induced release, temperature sensing, and cell imaging [3]. The Carbon-Dots in particular served as fluorescence imaging agents; importantly, they retained their photophysical properties throughout the preparation procedure of the hybrid assembly as well as after delivery to the cell targets.

There has been a growing interest in coupling Carbon-Dots and *metal nanoparticles (NPs)*, combining both the *luminescence* properties of the Carbon-Dots with the *plasmonic* properties of metal NPs (particularly noble metals such as gold and silver). Specifically, such composite particles could have useful biomedical properties, with the Carbon-Dots providing imaging capabilities and the metal NPs serving as conduits for therapeutic applications such as photothermal therapy (e.g., localized heating via light absorbance through the plasmonic bands of the metal NPs). Figure 8.4 depicts composite Carbon-Dot/AgNPs synthesized by C. Liu and colleagues at the Chinese Academy of Sciences, in which the Carbon-Dots were attached through electrostatic attraction onto the surface of a *silica layer* encasing the AgNPs [4]. The silica coating had a dual role, both immobilizing the Carbon-Dots thus preventing self-quenching, as well as maintaining separation *between* the Carbon-Dots and AgNP cores. This spacing enabled on the one hand occurrence of energy transfer between the two nanoparticle species, on the other hand maintaining sufficient distance to prevent complete quenching of the Carbon-Dots' fluorescence by the AgNPs.

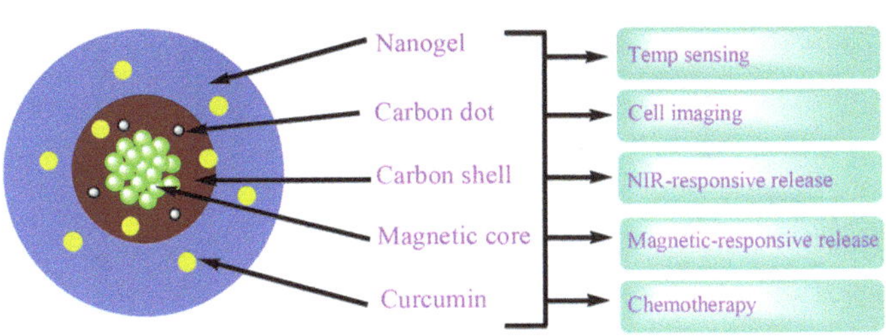

Fig. 8.3 Carbon-dots as components in a *theranostic* platform. The Carbon-Dots were used for fluorescence imaging of the localization of the theranostic particles. Reproduced from Wang et al. [3], with permission of the Royal Society of Chemistry

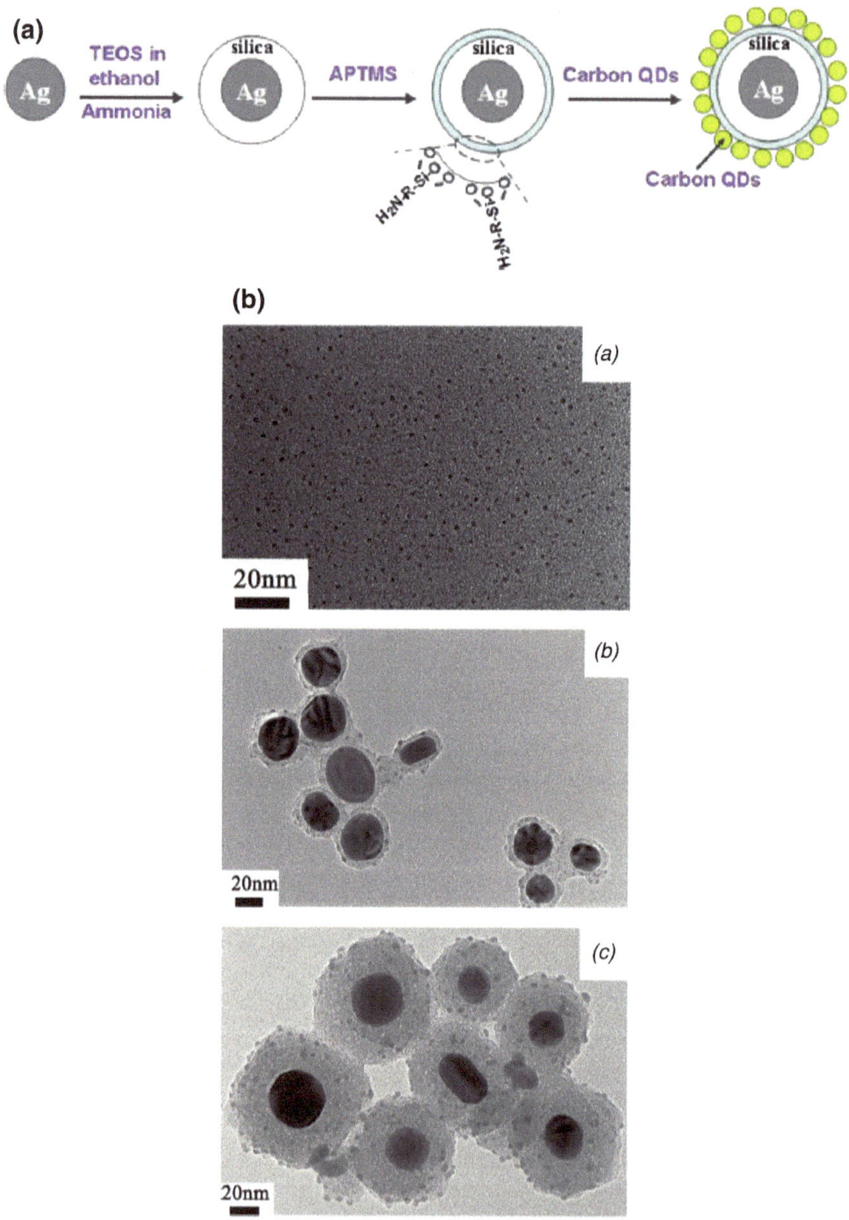

Fig. 8.4 Carbon-dot/silver nanoparticle composites. **a** Synthetic route; AgNPs are first coated with a silica layer, designed to separate the silver NP core and Carbon-Dot satellites, and the Carbon-Dots are electrostatically anchored onto the silica layer functionalized with positively charged moieties. **b** TEM images showing the Carbon-Dots (*a*), and Carbon-Dot/AgNPs with silica layers exhibiting different thicknesses (*b*, *c*). Reproduced from Li et al. [4], with permission of the Royal Society of Chemistry

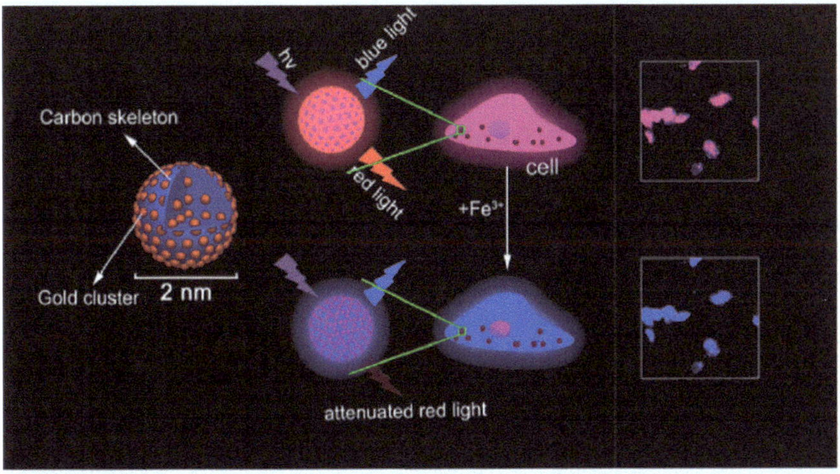

Fig. 8.5 Carbon-dot/gold cluster composite nanoparticles. The NPs luminesce both in *blue* and in *red*; quenching of only the red luminescence by Fe^{3+} enables ratiometric intracellular determination of Fe^{3+} levels. Reprinted with permission from Zhang et al. [5]. Copyright (2016) American Chemical Society

Composite Carbon-Dot/metal NPs combining the photophysical properties of the two particle types have been reported. Figure 8.5 depicts "core-shell" Carbon-Dot/Au nanocluster (NC) assembly which exhibited *dual* fluorescence emissions induced by single excitation wavelength [5]. Interestingly, the composite system, synthesized by X. Liu and colleagues at the Shanghai Institute of Ceramics, China, comprised of a 2-nm Carbon-Dot core coated with tiny Au NCs. This morphology presumably formed through a kinetic process in which the growth of the carbon core was constrained by nucleating Au clusters assembled upon its surface. Following the chemical reduction of gold with glutathione (GSH), the composite Carbon-Dot/Au nanoparticles emitted (upon UV excitation) both *blue* light (the Carbon-Dot fluorescence emission) as well as *red* (Au NC emission). The dual emission enabled, for example, two-color fluorescence microscopic cell imaging. Moreover, the researchers demonstrated that Fe^{3+} ions significantly attenuated the red fluorescence emitted by the Au NCs, making possible monitoring of iron levels in cellular environments through *ratiometric* analysis (i.e., recording the ratios between the red/blue emissions).

Carbon-Dot- or GQD-based composite materials for *nonbiological* applications have been also reported. Figure 8.6 schematically depicts a stretchable *photodetector* comprising a graphene layer on which GQDs were deposited [6]. In general, the prominent parameters determining the performance of photodetectors are the efficiency of *light absorbance* and generated *charge carriers*. Graphene is an excellent conductor, however it exhibits poor light absorption properties. To overcome this barrier, Y.F. Chen and colleagues at the National Taiwan University

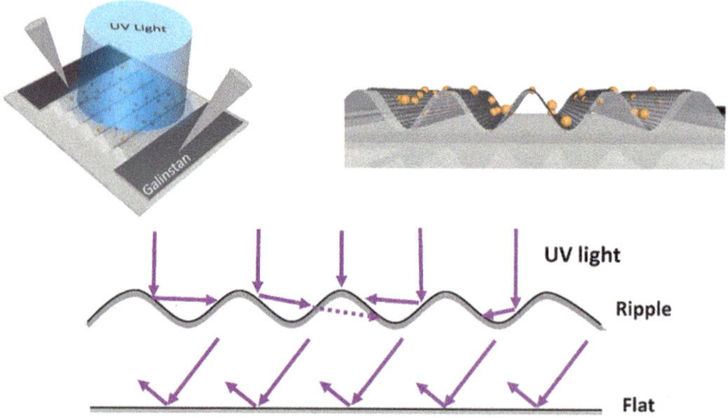

Fig. 8.6 Graphene/graphene-quantum-dot photodetector. The photodetector device (*top left*) comprises of a wrinkled graphene film with surface-deposited GQDs (*top right*). The "rippled" surface morphology induces reflections of the illuminating light, producing greater absorbance by the surface-deposited GQDs. Reprinted with permission from Chiang et al. [6]. Copyright (2016) American Chemical Society

have exploited the broad photon absorption capabilities of GQDs to create a sensitive photodetector through the coupled graphene-GQD system. The "wrinkled" film shown in Fig. 8.6 bestowed an enhanced light sensitivity as the incoming photons were reflected between the ridges upon the detector surface, absorbed by the surface-deposited GQDs, and generating a higher abundance of charge carriers within the graphene sheath.

As is the case in many hybrid materials, *synergistic effects* among the individual components have been observed in Carbon-Dot-containing composite materials. Figure 8.7, for example, illustrates a Carbon-Dot-containing *hydrogel* for which synergism was observed. Hydrogels have attracted interest as useful host matrixes due to their large internal surface area and molecular storage capacity. J.W. Steed and colleagues at the University of Durham, UK, for example, fabricated a Carbon-Dot/hydrogel hybrid from 5-aminosalycilic acid as the gel scaffold [7]. An interesting synergistic effect was observed in the system as incorporation of the Carbon-Dots appeared to have improved the *gelation properties* of the hydrogel—specifically increasing the gelation rate, consequently generating a stable and resilient gel. Moreover, the researchers reported that the *fluorescence emission* of the hydrogel-embedded Carbon-Dots was significantly higher compared to solution [by almost an order of magnitude]. This effect has not been fully clarified (as is the case with other photophysical aspects of Carbon-Dot luminescence), and was ascribed to the hydrophobicity of the hydrogel surfaces which presumably reduced solvent-induced quenching of the Carbon-Dots' fluorescence. The combination of extended hydrogels' surface area and enhanced Carbon-Dots fluorescence could be exploited for varied sensing applications.

Conceptually similar to the hydrogel-Carbon-Dot assembly depicted above, other composite materials comprised of Carbon-Dots that were organized in two or

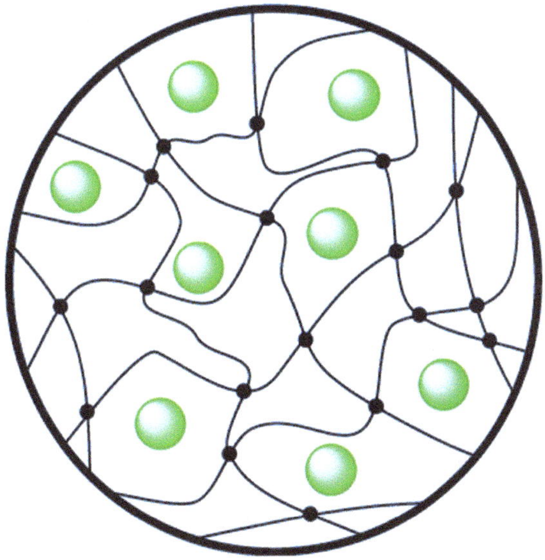

Fig. 8.7 Carbon-dots embedded within a hydrogel matrix. The schematic drawing underscores the physical separation between the Carbon-Dots (*green spheres*)

three dimensions through the use of inorganic *templates*. Figure 8.8, for example, illustrates a "honeycomb" architecture of Carbon-Dots, fabricated through the use of silica nanospheres as templating agents. The ordered Carbon-Dots were constructed by L. Qu and colleagues at Beijing Institute of Technology, China, through preassembly of silica colloids in a periodic structure, followed by electro-deposition of the Carbon-Dots [8]. A high-temperature annealing process was then carried out to produce a uniform Carbon-Dot layer upon the silica colloids. Notably, removal of the silica nanoparticle template generated a Carbon-Dot "sacrificial template" that could be used as a "mold" for further assembly of other functional nanostructures; the researchers have shown, for example, that gold NPs could be accommodated within the Carbon-Dot template, yielding an excellent substrate for surface-enhanced Raman scattering (SERS) sensing applications.

Other composite silica/Carbon-Dot materials have been shown to exhibit interesting optical properties. H.T. Chang and colleagues at National Taiwan University used a silane-containing derivative as a precursor material to prepare silicon-Carbon-Dots (SiC-dots) which were coupled to SiO_2 through formation of Si–O–Si and Si–O–C bonds [9]. Interestingly, SiC-dot/SiO_2 composites exhibiting different colors were fabricated through modulating the temperature and/or duration of the hydrothermal reaction (Fig. 8.9). The Carbon-Dot composites were employed for preparing different *glasses* having remarkable colors, both under conventional lighting as well as upon illumination at specific wavelengths. The SiC-dot/SiO_2 powders were resilient and stable, and could be further employed as *ink* material.

Composites of Carbon-Dots together with *conductive polymers* have been reported, exhibiting interesting electrical conductivity properties [10]. Figure 8.10 depicts a procedure for preparation of a Carbon-Dot/polypyrrole (PPy) composite

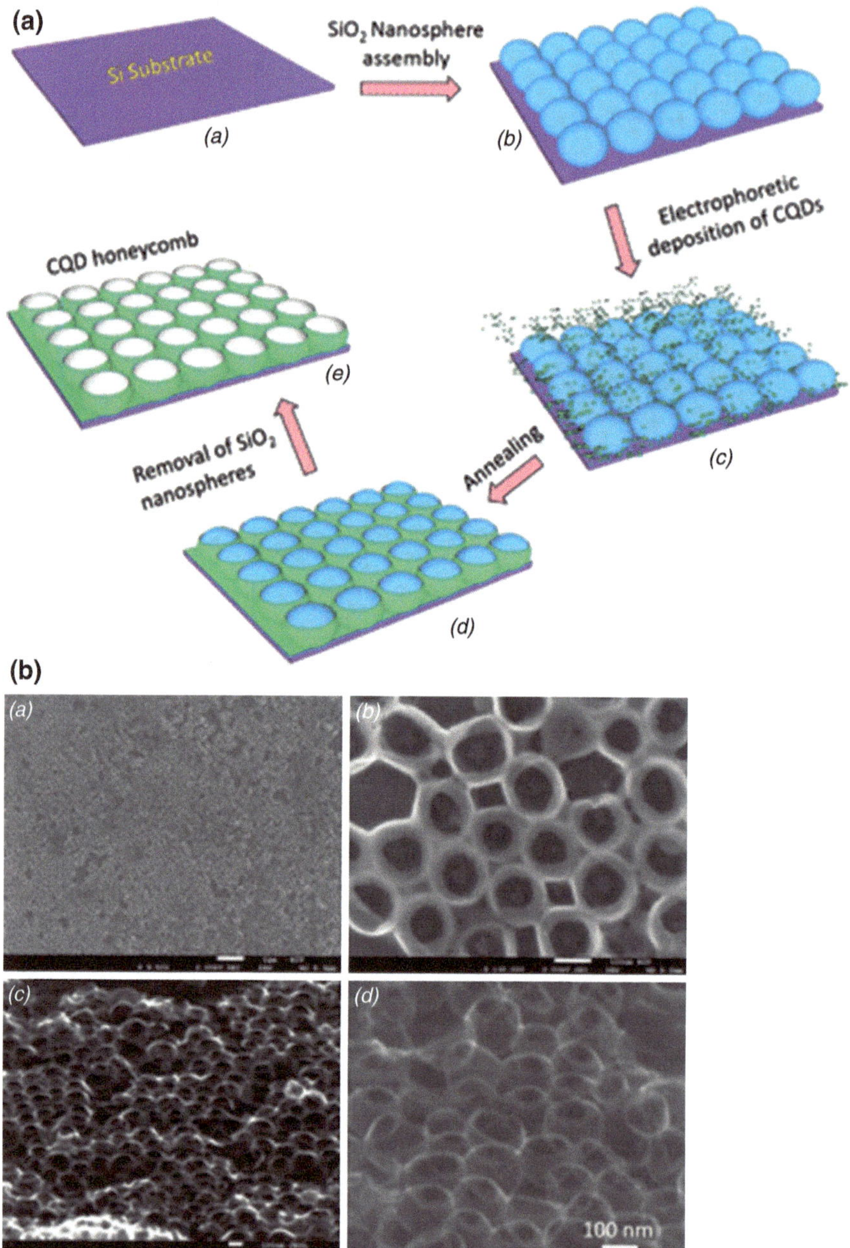
(a)
Si Substrate
(a)
SiO2 Nanosphere assembly
(b)
Electrophoretic deposition of CQDs
(c)
Annealing
(d)
Removal of SiO2 nanospheres
CQD honeycomb
(e)
(b)
(a)
(b)
(c)
(d)
100 nm

◀ **Fig. 8.8** "Honeycomb" carbon-dot architecture as two-dimensional support for Au NP deposition. **a** Assembly scheme. The Carbon-Dots are deposited upon a two-dimensional array of silica particles. An ordered Carbon-Dot architecture is generated following removal of the silica substrate. **b** SEM images of the Carbon-Dot honeycomb film. (*a*, *b*) *Top view*. Scale bar is 1 mm in (*a*) and 100 nm in (*b*). (*c*, *d*) images of a tilted film; scale bars correspond to 100 nm. Reproduced from Fan et al. [8], with permission of the Royal Society of Chemistry

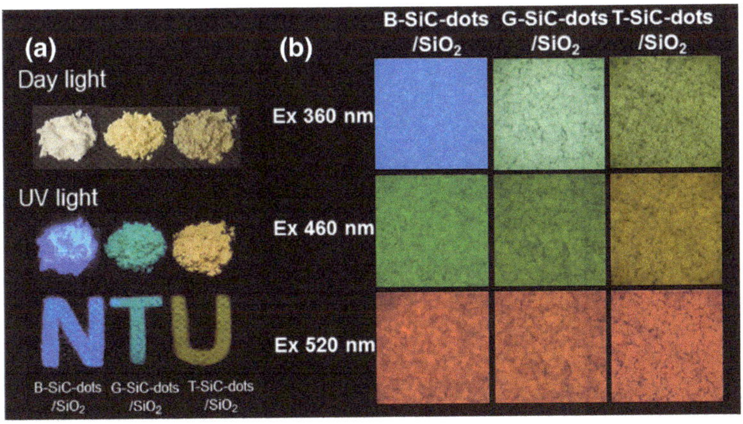

Fig. 8.9 Glassy materials comprising silicon-carbon dots (SiC-dots) coupled to silica. **a** photographs of glass powders of SiC-dots exhibiting different colors [*blue* (B-SiC-dots), *green* (G-SiC-dots), and *tan* (T-SiC-dots)], coupled to SiO_2, producing different colors upon irradiation with regular light and UV light. **b** Films comprising SiC-dots/SiO_2 composites illuminated at different wavelengths. Reprinted with permission from Shih et al. [9]. Copyright (2015) American Chemical Society

film and its conductivity profile. PPy is a well-known conductive polymer; interestingly, A. Chattopadhyay and colleagues at the Indian Institute of Technology, Guwahati, discovered that incorporation of the Carbon-Dots within the PPy matrix gave rise to significantly higher conductivity compared to the individual constituents alone (Carbon-Dots or PPy). The conductivity enhancement was ascribed to the *electron acceptor* properties of the Carbon-Dots (synthesized from citric acid and ethylene-diamine), which likely promoted hole mobility along the PPy backbone (PPy being an n-type semiconductor, i.e., an electron donor).

Other Carbon-Dot-containing composites have been shown to exhibit *electrical* properties. Figure 8.11 depicts *nanoribbons* constructed through mixing Carbon-Dots with styrene/styrenesulfonic acid monomers. In this study, M. Wang and colleagues at the Institute of Plasma Physics, China, have shown that the Carbon-Dots had an essential role in formation of the extremely long supramolecular nano-ribbons (reaching hundreds of micron in length), likely stabilized through electrostatic interactions between the positively charged Carbon-Dots and polystyrene-4-styrenesulfonate (PS-PSS) *oligomers* [11]. Interestingly, the Carbon-Dot/PS-PSS nanoribbons exhibited electrical conductivity that was facilitated through the PSS conductive network (as PSS is the most

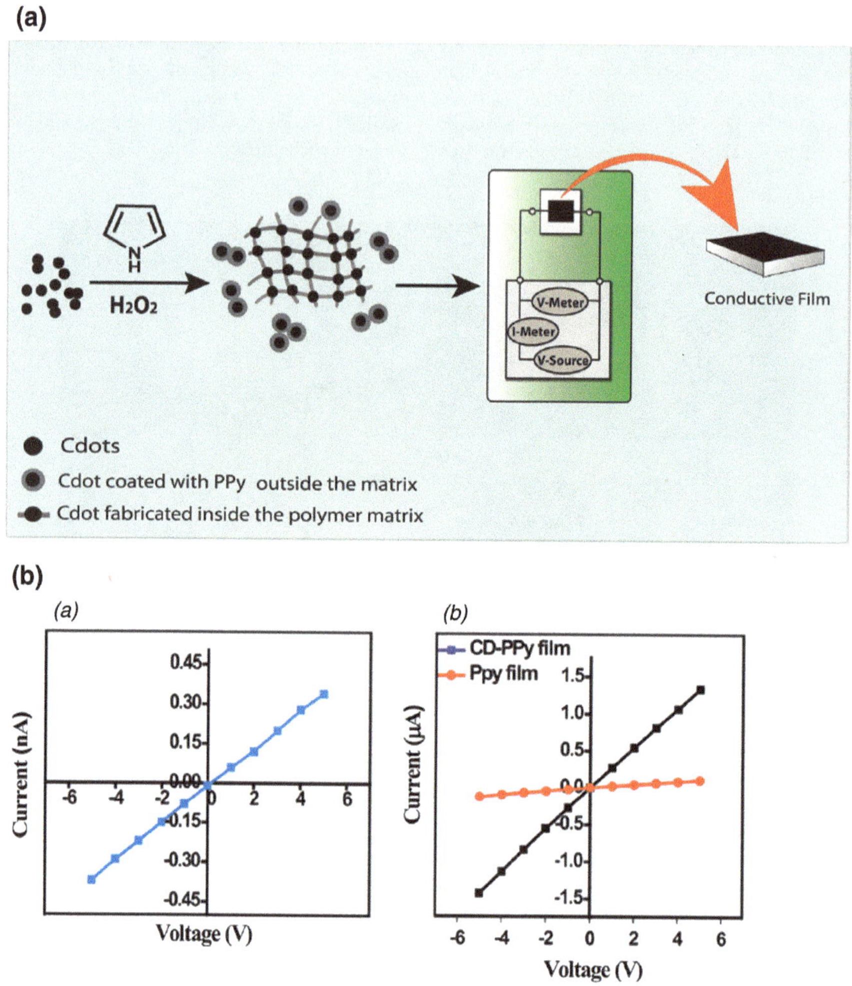

Fig. 8.10 Carbon-dot/conductive polymer composite. **a** Synthesis scheme: the polypyrrole (PPy)—Carbon-Dot matrix is assembled through mixing preprepared Carbon-Dots and pyrrole in the presence of hydrogen peroxide. The resultant composite material comprises Carbon-Dots both within and outside of the PPy framework. **b** Current–voltage (I–V) curves recorded for Carbon-Dot film (*a*), PPy and Carbon-Dot/PPy film (*b*). Note the orders-of-magnitude greater current generated in the Carbon-Dot/PPy film compared to the other compositions. Reprinted with permission from Pal et al. [10]. Copyright (2016) American Chemical Society

widely used organic semiconductor material). The embedded Carbon-Dots also made possible *visualization* of the nanoribbons through the multicolor luminescence of the dots.

Interesting application of Carbon-Dot composites for *printing* has been proposed in a recent study demonstrating that Carbon-Dots could be covalently coupled to *cellulose nanofibrils* (CNFs) which are common *paper* constituents [12]. The

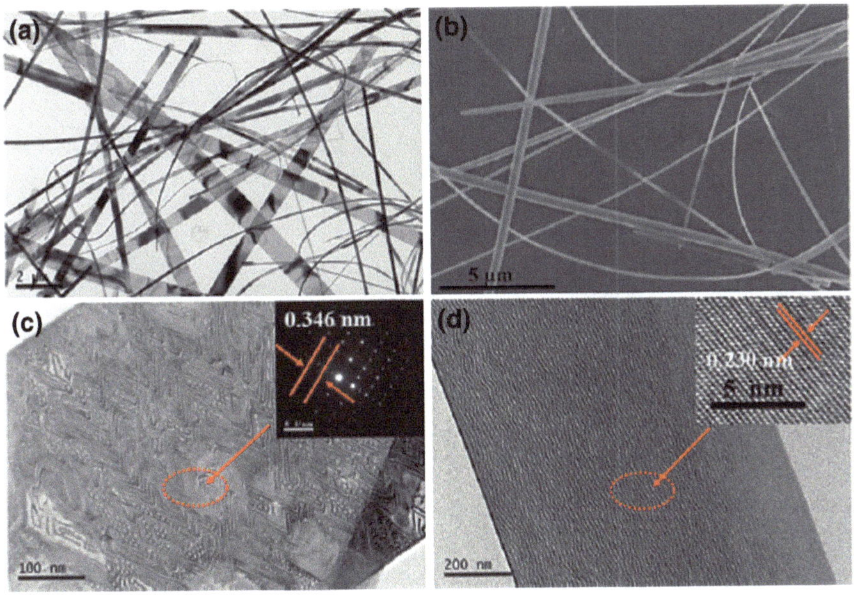

Fig. 8.11 Electron microscopy images of carbon-dot/polystyrene-4-styrenesulfonate (PS-PSS) nanoribbons. TEM (**a**) and SEM (**b**) images showing the extremely long Carbon-Dot/PS-PSS nanoribbons. **c**, **d** High resolution TEM (HR-TEM) revealing the crystalline graphite within the nanoribbons; the inset in (**c**) shows the selected area electron diffraction (SAED) pattern confirming the graphite crystal spacing, while the inset in (**d**) reveals the lattice fringes corresponding to the graphitic core of the Carbon-Dots. Reproduced from Zhang et al. [11], with permission of the Royal Society of Chemistry

simple synthesis procedure, developed by O.J. Rojas and colleagues at Aalto University, Finland, exploited the electrostatic attraction and subsequent covalent attachment of amine-containing Carbon-Dots to the carboxylate residues displayed upon *carboxymethylated* CNFs (CM-CNFs, Fig. 8.12). The Carbon-Dots retained their luminescence even after coupling to the nanofibers, opening the way for production of a "tunable luminescent paper." Indeed, the researchers fabricated thin films of the Carbon-Dot/CNF conjugate, which were both transparent and fluorescent (Fig. 8.12b).

As mentioned above, Carbon-Dots associated with metallic nanoparticles (NPs) have featured interesting optical and physical properties. Such systems exploit coupling between the fluorescence properties of the Carbon-Dots and plasmonic properties of the metals to achieve useful functionalities, such as sensing applications. Figure 8.13 illustrates spontaneous formation of Carbon-Dot/AgNPs through incubation of Carbon-Dots with a common silver salt in aqueous solution. Specifically, J. Wang and colleagues at Northeastern University, China, have shown that a very short (5-min) incubation of the Carbon-Dots with Ag^+ ions at 50 °C resulted in accumulation of AgNPs upon the Carbon-Dots' surface [13]. The

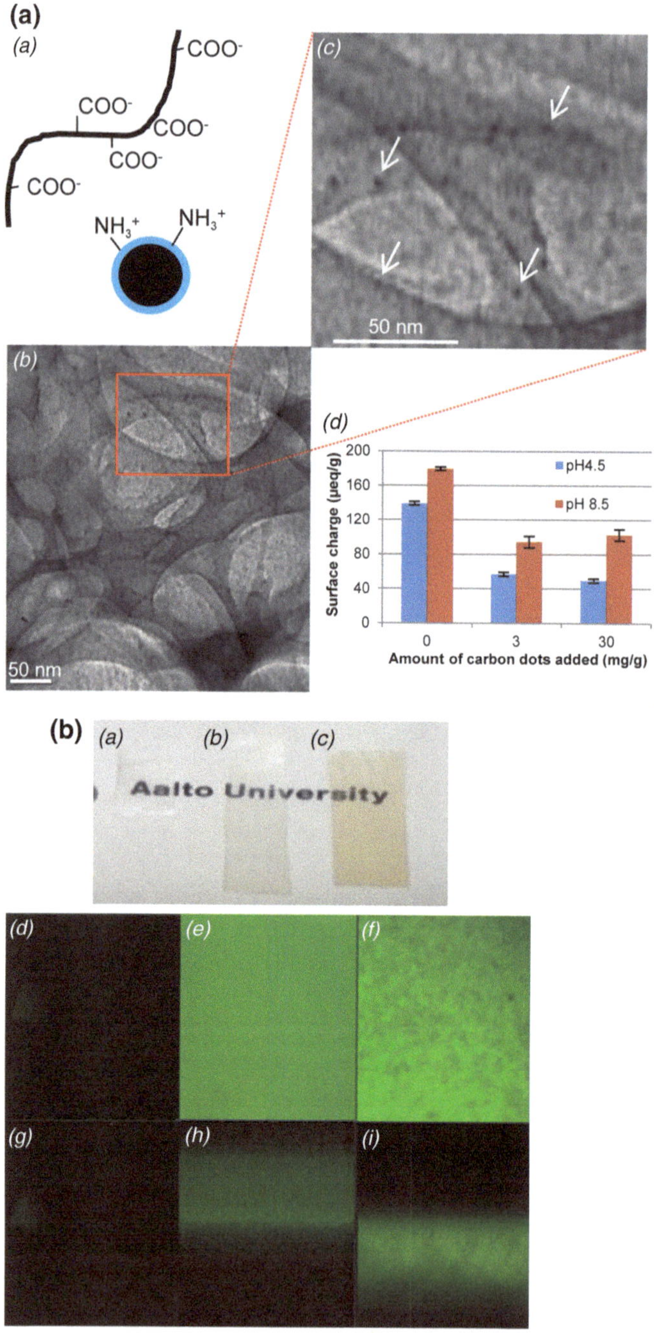

(a)
(a)
COO-
COO-
COO-
COO-
COO-
NH3+
NH3+
(c)
50 nm
(b)
50 nm
(d)
Surface charge (μeq/g)
200
160
120
80
40
0
pH4.5
pH 8.5
0
3
30
Amount of carbon dots added (mg/g)
(b)
(a)
(b)
(c)
Aalto University
(d)
(e)
(f)
(g)
(h)
(i)

◀ **Fig. 8.12** Carbon-dot/cellulose for printing applications. **a** Characterization of the Carbon-Dot/cellulose conjugate. (*a*) the Carbon-Dots are attracted to the cellulose framework through electrostatic attraction; (*b*, *c*) TEM images showing the Carbon-Dots attached onto the cellulose matrix surface; (*d*) evolution of the cellulose surface charge upon Carbon-Dot binding. Higher Carbon-Dot concentration results in lower surface charge. **b** Optical properties of the films. *Top* Camera images of films comprising (a) CM-CNF, (b) CD-coated CM-CNF, and (c) CNF-CD. *Bottom* confocal microscopy images of the films' surface (d–f) and cross-section (g–i). Images (d) and (g) correspond to CM-CNF film, (e) and (h) are the CD-coated CM-CNF films, and (f) and (i) are the CNF-CD films. Best staining with Carbon-Dots observed in case of the CM-CNF film. Reprinted with permission from Junka et al. [12]. Copyright (2014) American Chemical Society

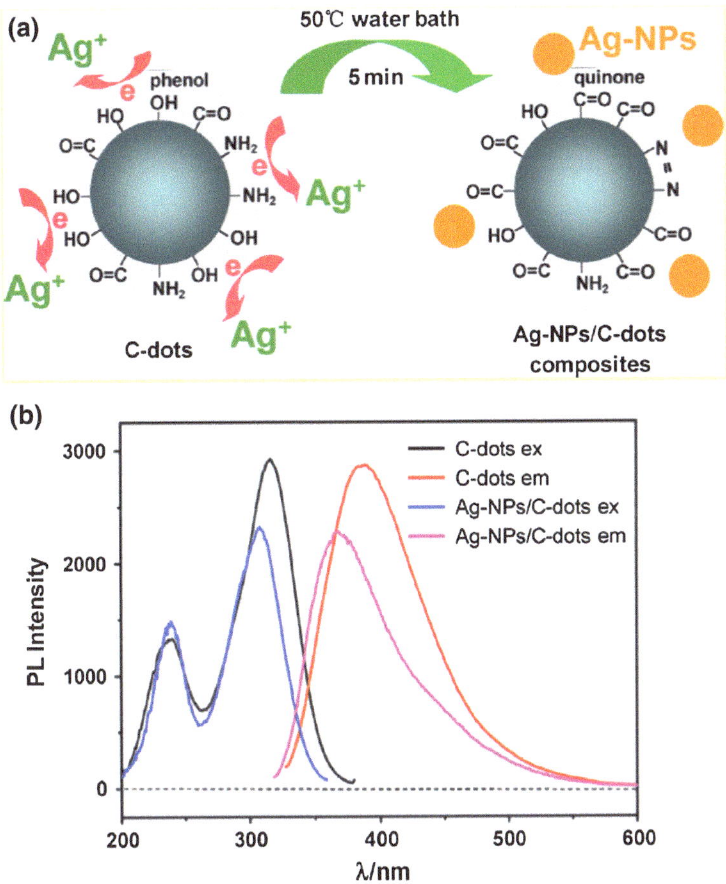

Fig. 8.13 Carbon-dot/silver nanoparticles. **a** Cartoon showing the growth of the AgNPs upon the Carbon-Dot's surface through reduction of the Ag^+ ions by Carbon-Dots' surface residues. **b** Modulation of the intensities and shifts of the excitation/emission signals of Carbon-Dots in the Carbon-Dot/AgNP composites. Reprinted with permission from Shen et al. [13]. Copyright (2013) American Chemical Society

growth of AgNPs was ascribed to reduction of Ag^+ by electron-donating groups at the Carbon-Dots' surface, particularly hydroxyl- and amine-containing moieties (Fig. 8.13a). The hybrid Carbon-Dot/AgNPs exhibited significant modulation of both the visible spectrum corresponding to the AgNPs, as well as the fluorescence excitation/emission of the Carbon-Dots (Fig. 8.13b). This spectral modulation could be employed, in fact, as a vehicle for sensing Ag^+ ions in solution.

References

1. Y. Liu, Y. Tian, Y. Tian, Y. Wang, W. Yang, Carbon-dot-based nanosensors for the detection of intracellular Redox State. Adv. Mater. **27**, 7156–7160 (2015)
2. L. Zhou, Z. Li, Z. Liu, J. Ren, X. Qu, Langmuir **29**, 6396–6403 (2013)
3. H. Wang, J. Yi, S. Mukherjee, P. Banerjee, S. Zhou, Nanoscale **6**, 13001–13011 (2014)
4. J. Li, B. Zhang, F. Wang, C.-Y. Liu, New J. Chem. **35**, 554–557 (2011)
5. L. Zhang, D. Wang, H. Huang, L. Liu, Y. Zhou, X. Xia, K. Deng, X. Liu, ACS Appl. Mater. Interfaces **8**, 6646–6655 (2016)
6. C.-W. Chiang, G. Haider, W.-C. Tan, Y.-R. Liou, Y.-C. Lai, R. Ravindranath, H.-T. Chang, Y.-F. Chen, ACS Appl. Mater. Interfaces **8**, 466–471 (2016)
7. A. Cayuela, S.R. Kennedy, M.L. Soriano, C.D. Jones, M. Valcarcel, J.W. Steed, Chem. Sci **6**, 6139–6146 (2015)
8. Y. Fan, H. Cheng, C. Zhou, X. Xie, Y. Liu, L. Dai, J. Zhang, L. Qu, Nanoscale **4**, 1776–1781 (2012)
9. C.-C. Shih, P.-C. Chen, G.-L. Lin, C.-W. Wang, H.-T. Chang, ACS Nano **9**, 312–319 (2015)
10. A. Pal, M.P. Sk, A. Chattopadhyay, ACS Appl. Mater. Interfaces **8**, 5758–5762 (2016)
11. G. Zhang, M. Yan, X. Teng, H. Bi, Y. Han, M. Tian, M. Wang, Chem. Commun. **50**, 10244–10247 (2014)
12. K. Junka, J. Guo, I. Filpponen, J. Laine, O.J. Rojas, Biomacromolecules **15**, 876–881 (2014)
13. L. Shen, M. Chen, L. Hu, X. Chen, J. Wang, Langmuir **29**, 16135–16140 (2013)

Chapter 9
Conclusions and Future Outlook

For a young and emerging field, Carbon-Dots (and their close relatives graphene quantum dots) have already made a significant impact in diverse scientific disciplines. Carbon-Dots have opened new avenues in biological imaging, photonics, electro-optics, and others. In particular, the unique luminescence properties of Carbon-Dots, primarily the excitation-dependent emission phenomena, have been a prominent thread in this field, making possible diverse experiments and analytical applications developed in the past several years. The optical and spectroscopic properties of Carbon-Dots will likely continue to shape the future development of this field.

This Book summarizes the evolution and current state of Carbon-Dot research. In particular, a recurring motif in many of the published studies is the close relationship between synthesis routes and the photophysical properties of Carbon-Dots. Indeed, proliferating means of controlling the chemical features of Carbon-Dots, particularly as related to *surface functionalities*, has enabled fine tuning of Carbon-Dots' luminescence properties. This capability will likely continue to drive further progress in this field, as sophisticated synthetic acumen will produce Carbon-Dot systems exhibiting predesigned optical features. As presented in numerous publications, chemical control of Carbon-Dots' luminescence has likely been among the most dramatic facet in Carbon-Dot research and technological development.

Perhaps naturally, Carbon-Dot research still faces challenges, both conceptual as well as technical. Elucidation of the precise physical mechanisms responsible for the unique spectroscopic properties of Carbon-Dots is still lacking. Indeed, while significant progress has been made since inception of the field toward establishing a firm molecular basis for Carbon-Dots' photophysical properties, putting together all the "puzzle-pieces" is yet to be completed. In particular, experimental and theoretical work needs to confirm the prominence of the "surface energy states" as the determinant for Carbon-Dots' luminescence, and more investigations are necessary to shed light on the relationship between Carbon-Dots' surface chemistry and optical properties of the particles. It is anticipated that further studies will contribute to better understanding of these aspects in coming years.

© Springer International Publishing Switzerland 2017

R. Jelinek, *Carbon Quantum Dots*, Carbon Nanostructures,
DOI 10.1007/978-3-319-43911-2_9

Other challenges are more technical in nature. Due to the centrality of Carbon-Dots' optical properties, an intensive research currently aims at improving the luminescence characteristics of Carbon-Dots. Specifically, in many instances the *luminescence intensity* and related *quantum yields* are still insufficient for many practical applications. Similarly, significant efforts have been, and will continue to be directed toward *tuning* Carbon-Dot colors, as well as reducing/expanding the excitation-dependent luminescence range. As indicated above, such objectives will likely involve careful modulation of Carbon-Dot synthesis and functionalization.

Biological applications of Carbon-Dots will likely continue to be at the forefront of research and development efforts in this field. While the attractiveness of Carbon-Dots in biology is obvious—as these nanoparticles are in many instances produced from harmless natural sources—further research will have to fully establish the biological profiles of these nanoparticles. In particular, solid experimental evidence is necessary to evaluate Carbon-Dots' toxicity and to probe the relationship between biological activity and Carbon-Dots' sizes and surface functional groups. Indeed, broader applicability and commercialization of Carbon-Dots in biology and other disciplines will have to go hand in hand with comprehensive analyses of their physiological impacts. In that regard, the evolution of *inorganic* quantum dots and their practical applications will continue to provide lessons and directions.

In conclusion, Carbon-Dots—practically "discovered" less than a decade ago—constitute (another) fine example of the wonders of the "carbon nanoworld" in particular, chemical research in general. Similar to other colloidal systems in which the nanoscale dimensions have opened new realms for research and applications, Carbon-Dots have already bestowed their fair share of new, interesting phenomena and scientific excitement. The inter- and multidisciplinary nature of Carbon-Dot research will likely continue to inspire scientists, and generate new systems, experiments, and applications.

FSC
www.fsc.org
MIX
Papier | Fördert
gute Waldnutzung
FSC® C083411

Kolibri 360 GmbH
Ferdinand-Jühlke-Straße 7
99095 Erfurt, Deutschland
produktsicherheit@kolibri360.de